Animal Electricity

HOW WE LEARNED THAT THE BODY
AND BRAIN ARE ELECTRIC MACHINES

Robert B. Campenot

 Harvard University Press   Cambridge, Massachusetts · London, England   2016

First printing

Library of Congress Cataloging-in-Publication Data
Campenot, Robert B., 1946–
Animal electricity : how we learned that the body and brain are
electric machines / Robert B. Campenot.
          pages cm
Includes bibliographical references and index.
ISBN 978-0-674-73681-8
1. Electrophysiology.   2. Biophysics.   3. Electricity—
Physiological effect.   I. Title.
QP341.C36 2016
612.8'13—dc23
2015014359

For Mary Kay and Jeff

CONTENTS

PREFACE

This book has had a long incubation. I can trace its origins back to 1968 when I graduated with a B.A. in psychology from Rutgers University and entered the graduate program in physiology at UCLA. Having not majored in biology, I found myself unprepared for some of the courses in physiology. I remember struggling with the Nernst equation, which provides a means to calculate the voltage that develops across the cell membrane from unequal concentrations of the ions potassium, sodium, and chloride on each side of the membrane. The voltage is produced by ions flowing across the membrane, but the concentrations of the ions never changed in the examples that the class was given to work through. I asked the professor why, and he said that the changes in concentrations were very small and could be ignored. Fine, I could now do the calculations, but I felt there was something I needed to understand that was missing. During my attempt to satisfy this feeling, I discovered *The Feynman Lectures on Physics* buried on a bottom shelf in the stacks of the UCLA library. The three large, red volumes were nonetheless hard to miss, and while seated cross-legged on the floor with volume 2 on my lap I came upon a quotation that radically changed my understanding of the physical world: "If you were standing at arm's length from someone and each of you had *one percent* more electrons than protons, the repelling force would be incredible. How great? Enough to lift the Empire State Building? No! To lift Mount Everest? No! The repulsion would be enough to lift a 'weight' equal to that of the entire earth!" (Feynman, Leighton, and Sands 1964, 1.1). Richard Feynman went on to explain that, like all objects, we are made of very nearly equal numbers of protons and electrons finely mixed together, neutralizing each other's charges and allowing very little imbalance. Ultimately, I realized that the tiny electrical imbalances that do exist within our bodies are of enormous importance, giving rise to all sensation, movement, awareness, and thinking—most everything we associate with being alive. What was missing in the "explanation" given to me by the professor was the relationship of his answer to a principle of nature fundamental enough to be introduced on the first page of Feynman's book.

During the years since I have realized more and more that the imagery scientists carry around of the processes operating in the world has a tremendous impact on their thinking. This was evident in the history of neuroscience when many researchers thought that the release of chemicals across synapses would be too slow to play a role in communication between nerve cells. We now know that release of chemical neurotransmitters is the principal mechanism of synaptic transmission, which can occur at frequencies in the hundreds per second. When one fully appreciates the magnitude of the thermal motion of atoms and molecules on the nanoscale, fast chemical transmission seems perfectly reasonable.

The nerve impulses (action potentials) that travel along nerve fibers (axons) are often called spikes because of their sharp appearance on an oscilloscope screen or computer display, but action potentials can reach lengths of a third of a meter in fast-conducting motor axons that connect the spinal cord to skeletal muscle. This means that instead of a lineup of a great many narrow, sharp spikes marching along the axon, in reality the motor axons carry broad waves with only about one and a half waves occupying the axons extending the full length of your arm at any one time. This change in imagery makes a big difference in how one thinks about conduction of action potentials along axons; for example, it explains why the conduction velocity of the axon is increased in large-diameter axons and by the presence of electrically insulating myelin sheaths.

I have also learned that very few people, including a surprisingly small number of neuroscientists, actually have a reasonably accurate knowledge of how action potentials are produced and travel along axons and many of the other aspects of how the nervous system produces and utilizes electricity. This knowledge is seen as highly specialized, and in the era of molecular biology it tends to be relegated to background presented in introductory and mid-level courses and never considered again.

A few other realizations add to the stew of my motivations for writing this book. One is the lack of explanations for how things work in many science books aimed at a general audience. In my background reading I have come across many references to the Leyden jar, which was the first device invented that could store electric charge and played a fundamental role in the early history of research into the nature of electricity. In electrical parlance, the Leyden jar is a capacitor, but few sources adequately explain how capacitors work. It is even harder to find an adequate explanation of how electric eels and electric torpedo rays produce electric shocks. Also, there are frequent inadequacies in descriptions of experiments and their interpretation. My experience writing many scientific research papers has taught me that writing is the most rigorous form of thinking. One can perform a series of experiments aimed at elucidating the mechanism of a certain phenomenon, but, until written for publication, there is an increased chance that some flaw in experimental design and/or interpretation will be discovered that will change the conclusions. From the start of my career I have always enjoyed the challenge of writing clear and convincing scientific papers.

This is the background out of which my book has grown. The time to begin the book was provided by a sabbatical leave during the 2012–2013 academic year, and I have worked on it ever since. It is my attempt to provide an accurate appreciation for how conduction of action potentials and the other important phenomena of animal electricity work that is neither mathematical nor dumbed down. The explanations start with a description of how electricity is conducted in biological tissues, presented so that readers without backgrounds in biology, physics, or chemistry can understand it. From there the imagery is built up to provide an understanding of the essential processes that produce animal electricity. The writing is laced with history not only for human interest but also as a context for discussing the key experiments that produced the knowledge that our bodies and brains are electric machines. The book explains some of the medical applications arising from animal electricity, such as the electrocardiogram and the emerging field of brain-machine interfaces aimed at restoring movement to paralyzed individuals and developing bionic vision. In our world dominated by the Internet, opportunities for self-education are greater than ever before. It is my hope that my book will stimulate its readers to learn more and make the available knowledge of how the nervous system functions and the emerging opportunities for helping individuals with damaged nervous systems far more accessible to them.

INTRODUCTION

We are electric machines. This is true of all members of the animal kingdom. The discovery of animal electricity had a major influence on the shift from supernatural to mechanistic explanations of the functions of the body and brain—one of the most significant transformations of human thought that has ever occurred. The activities of our sensory systems, nervous system, and muscle all run on the battery power of cell membranes. Without a source of electric power for these systems we would be less than vegetables, unaware of ourselves, or our environment, without thoughts, and incapable of any meaningful response. Yet it is a rare individual in our enlightened and educated population that has even a rudimentary knowledge of the electrical functions of our bodies and brains. The focus on molecular biology in today's college and university biology programs has relegated the understanding of animal electricity to a small subset of specialists.

It seems a shame to dismiss the vast majority of interested people to fundamental ignorance about how our bodies and brains work. Investigations of electricity captivated the public in the eighteenth century when nobody had any idea how decisions made in our brains could translate into the movement of our muscles, much less how our brains make decisions in the first place. How nerves transmit signals is now well understood, but, alas, by only a very few. However, our effectiveness in dealing with our world depends to some extent upon our knowledge of how it really works, and the more accurate and complete that knowledge is, the better able we are to understand what is happening to us and how to deal with it. For example, several years ago during a visit to my laboratory, a person in a wheelchair because of a spinal cord injury asked me when I thought treatment of spinal cord injury would be advanced enough so that someone paralyzed after breaking his neck while playing basketball could expect to be back on the court playing with his team the next season.

Damaging the nervous system will always be very serious. On several occasions I have witnessed the anger that people with disabling neurological injuries feel toward medical professionals who are unable to offer them a cure. The anger extends to the scientific research community, which they feel should have provided cures by now were it not for researchers pursuing research to satisfy their own curiosity rather than aiming for cures. Readers of this book will gain an appreciation of the intricate workings of the cells that make up our brains, nerves, and muscle. They will realize that a properly functioning brain and spinal cord are precious things worthy of our tender loving care. If treatments that make possible substantial repair to damaged spinal cords or even brains are ever developed, such treatments will be among the great scientific achievements of the, hopefully twenty-first, century. While diseases including devastating ones such as Alzheimer's disease and Lou Gehrig's disease may one day be largely preventable, accidents will happen, and damage to the brain and spinal cord that cannot be repaired will always be with us. Readers will learn about bionic prostheses under development that interface with the brain and may significantly improve the independence and quality of life of those with permanent damage to their nervous systems.

Medical treatments and other practical benefits are just part of the reason for discovering how our world and the creatures that populate it work. It is difficult to separate existing knowledge into well-defined categories, and nearly impossible before the knowledge is discovered. This seems so obvious, but a surprising number of people appear to believe that it should be possible to tell in advance whether a discovery that has not been made yet will cure a medical condition or be of some other practical value. In fact, narrowly targeted research often misses the mark by miles. Fortunately for us, the curiosity of our predecessors has been a major driver of research and has provided a wealth of information upon which we now draw. For example, the reader will learn that two fish that produce electric shocks, the electric eel of South American rivers and the widely distributed oceanic torpedo ray, were major subjects of study in the eighteenth century and helped reveal the nature of electricity. In fact, the electric organ of the Mediterranean torpedo ray provided an example from nature, which guided Alessandro Volta to invent the first battery. In the 1940s Alan Hodgkin and Andrew Huxley of Cambridge University discovered the actual mechanism of nerve impulse conduction in nerve fibers through experiments with the giant nerve fiber of the squid. It was only because of the previous curiosity-driven research by the zoologist J. Z. Young that it was known that squids have giant nerve fibers. Someone to whom I had described my research with nerve cells cultured from rat tissue once asked whether it is possible to learn anything about the function of human nerves from experiments with rat nerves. My hope is that readers of this book will discover the answers to this and many other questions.

The knowledge that was needed to discover the mechanisms of animal electricity started with the fundamental concept that everything is made of atoms, which harbor within their structure electrically charged particles—electrons and protons. It is unfair to assume that readers who have not had a physics or chemistry course since high school decades ago still have this basic knowledge. Therefore, the challenge was to supply the necessary scientific background in an engaging rather than a didactic, and therefore deadly dull, manner. The worst thing an author can do is make the reader feel that he or she is back in high school—baby boomers, remember the Paul Simon song! This would turn off readers who have long forgotten most of their high school chemistry and physics as well as readers who are scientists that know these subjects well enough not to need a didactic review. What is an author to do? My answer is to embed the review of the basics in interesting history because the most fascinating aspect of science to me is how we know what we know; for example, how did we ever learn about a world made of atoms and electrons that we can never see? It wasn't easy, and getting the method down so that real progress could begin took over 2,000 years. As an experimental scientist myself for over four decades, I see more in the historical experiments than I could have ever appreciated earlier in my career. So I think a historical treatment will captivate laymen and scientists alike.

One of the problems with books on biological subjects intended to be accessible to general readers is the tendency of biologists to present every detail and complexity. I have fought this by focusing on the most important and clear examples of important phenomena. For example, nerve cells, now usually called neurons, communicate with each other and with muscle by releasing chemicals known as neurotransmitters at contact points called synapses. In the discussion of the release of neurotransmitters at synapses I focus on the release of acetylcholine from motor nerve terminals. There are many other neurotransmitters in the brain and lots of variation in how they work, but I only discuss the important differences between the synapses in the brain, which are decision makers, and the synapses on muscle that function as amplifiers. Unlike textbooks, this book is not constructed to give the reader lists to memorize for a test. You don't need to see in print here the name of every neurotransmitter in existence. The goal is that the fundamentals gained by reading this book will help you better understand the information about the nervous system that you encounter in the future.

Imagery is the most important aspect of the book for capturing the reader's interest and providing real, not watered-down, understanding. I have thought about the subject matter in the book for many years, but I still had insights left to discover when I began writing. One of my favorites is that ion channels are really electric wires across the cell membrane. Potassium channels, for example, are lined with negatively charged proteins and contain mobile positive charges in the form of potassium ions that move through the channels. This is directly analogous to metal wires, which contain fixed positive charges in the form of protons and mobile negative charges in the form of electrons that carry current along the wires. This imagery is important because it sets up the reader to understand how the difference in potassium concentration and the electric field across the membrane both control the flow of electric current, which produces a voltage across the membrane.

Any book that purports to explain how electrical phenomena really work must have some mathematical formulas and diagrams. I have kept formulas at a minimum. There is only one the reader needs to understand, and that is the simplest—Ohm's law. There are a couple of others readers need to see and accept—the Nernst equation and the Goldman-Hodgkin-Katz equation—but the relationships in those equations are described with imagery so that readers can simply acknowledge those equations and move on without compromising their understanding at all. There are, necessarily, a few simple circuit diagrams, but they are intended as aids accompanying the explanations. The circuit diagrams should be helpful for many readers, but words alone should suffice for readers uncomfortable with circuit diagrams.

Animal electricity, from the story of its replacement of animal spirits, to the detailed understanding of how neurons and muscles work, the understanding of animals in nature, and attempts to provide treatments for the sick and injured, is a fascinating story that is central to our understanding of our world and ourselves. Many popular renditions of science have left me unsatisfied when the basic science is glossed over by the author as too complicated or too uninteresting for most readers. This should hardly be the case in a world where the opportunities for self-education are greater than ever before. Teachers at all levels must realize that learning must be lifelong to keep up with our fast-changing world. They must present knowledge worth knowing, in a form that makes it a pleasure to acquire and that can be comprehended with a minimum of detailed background. This book is my attempt to instill in readers an accurate understanding of animal electricity, the diversity of its expression throughout the animal kingdom, how it was discovered, how it works, how it can go wrong, and how it can be and is being utilized to improve our lives. Writing—the most intense and critical form of thinking—has provided me with some new insights into animal electricity. My hope is that some of these insights will enlighten even those readers already knowledgeable in the field.

The balance of this introduction consists of brief summaries for each chapter. The illustrations in the book that are not attributed to a source were created by me in Adobe Illustrator.

Chapter 1, "Animal Electricity," begins the book with a brief history of the observations and experiments that produced the transition from the animal spirits theory adopted during the time of Aristotle to the theory of animal electricity that developed alongside the emerging knowledge of electricity during the eighteenth century. Knowledge progresses by a process of explaining the unknown in terms of the familiar, and for two millennia all mechanisms that were invented to explain conduction in nerve and movement of muscle were essentially plumbing. The dominant theme involved a mysterious fluid called "animal spirits" residing in tanks (the ventricles) deep within the brain and conducted through the nerves to inflate and thereby contract the muscles. The schemes became more and more intricate, but there was no hope of figuring out what was really going on until electricity and atomic theory came on the scene. The process of discovery involved keen observers and experimentalists with the courage and stamina to believe their observations rather than what the authorities of the day were telling them.

With the stage set for modern electricity to enter the scene, Chapter 1 fast-forwards to the explanation of the different forms of electric current. This is done within a historical context that features development of the Leyden jar capacitor, Benjamin Franklin's designation of positive and negative charge in the eighteenth century, and the measurement of the charge of a single electron by Robert Millikan in 1909–1913. The concept is developed of electric current traveling in different forms through different materials—electrons in wires, ions in solutions, and as an electric field across a thin insulator like the cell membrane. Explanations of some of the units of electricity, Ohm's law, and the analysis of a simple circuit diagram are presented along the way. Chapter 1 ends with a discussion of the history of experiments into the electricity produced by electric eels and torpedo rays and the development of the concept of animal electricity that replaced animal spirits as the driving force for conduction in nerve and contraction of muscle.

Chapter 2, "A World of Cells, Molecules, and Atoms," lays out the biological playing field within which animal electricity is manifest. The historical context progresses from the ancient world where the powers of observation were limited to the naked eye, to the enhanced powers of the microscope, and ultimately to the world of molecules and atoms accessible only to the powers of experimentation. Two themes permeate this chapter. One theme is the scientific concepts and the instrumentation that produced the modern image of the world on the microscopic, molecular, and atomic levels. The other theme is the image of the world on those small scales where events that we tend to think of as slow from our experience, such as diffusion within fluids, can happen with lightning speed. The goal is for the reader to get down on that scale and feel what it must be like to be a molecule or an atom. Along the way several important concepts are introduced; for example, the perpetual (Brownian) motion of molecules and atoms, the electrically polarized (polar) nature of water molecules, the structure of the cell membrane in a watery environment, and how ions in a solution carry electric current. By the end of Chapter 2 the stage is set for the explanation of how cells make electricity, which is developed in Chapters 3, 4, and 5.

The cell membranes of all cells contain machinery that generates voltages between the inside and outside of the cell. The voltage difference develops across the cell membrane because of differences between the inside of the cell and the outside of the cell in the concentrations of electrically charged particles (ions), the most important being the positively charged ions potassium, which has a higher concentration inside the cell, and sodium, which has a higher concentration outside the cell. Protein channels in the cell membrane, which can be in open or closed configurations, control the permeability of the membrane of nerve cells (usually called neurons) to potassium and sodium. Potassium and sodium ions are both positively charged so the flow of potassium down its concentration gradient out of the cell leaves the inside of the cell with a negative charge, and the flow of sodium down its concentration gradient brings positive charge into the cell. The voltage developed across the membrane due to the flow of ions is known as the membrane potential. Thus, the cell membrane has two major batteries: the potassium battery with its negative terminal inside the cell, and the sodium battery with its positive terminal inside the cell. When open potassium channels dominate, the resulting negative membrane potential is called the resting potential; and when open sodium channels briefly dominate, the membrane potential moves in the positive direction, producing a nerve impulse (now called an action potential), which travels along a nerve fiber (now usually called an axon).

Chapter 3, "The Animal Battery," begins with a description of the mechanisms that produce and control the membrane potential. Three membrane proteins are featured: the sodium-potassium exchange pump, the voltage-gated potassium channel, and the voltage-gated sodium channel. The sodium-potassium exchange pump consumes energy to transfer sodium out of the cell and potassium into the cell to establish the concentration gradients that drive potassium and sodium through their respective channels to produce the membrane potential. The equilibrium potential for an ion is the voltage that the ion would produce across a membrane if the membrane were permeable to that ion alone. How multiple ions, all with different equilibrium potentials and different relative permeability, combine to produce the membrane potential is described. This is represented in a simple electric circuit model of the membrane of a neuron.

The basic principles presented in Chapter 3 provide the framework for discussing how the mechanism of the action potential was figured out, which is the subject of Chapter 4, "Hodgkin and Huxley before the War." The war was World War II, and Alan Hodgkin and Andrew Huxley were at Cambridge University. However, the story begins much earlier than that, with Jan Swammerdam (1637–1680), a Dutch biologist who pioneered an experimental system consisting of an isolated frog motor nerve with the muscle attached that is still used to this day. Hodgkin used the same system and started out stimulating the nerve electrically and measuring just the strength of contraction of the muscle. From his results Hodgkin figured out that a loop of electric current projects forward along the axon within the action potential. He eventually got hold of an oscilloscope that allowed him to directly observe the action potentials traveling simultaneously along many axons in the nerve and confirmed this conclusion. This loop of current, called the local circuit, plays a fundamental role in the conduction of the action potential along the axon. This work, which formed the basis of Hodgkin's undergraduate thesis at Cambridge, was published in the *Journal of Physiology*. It became clear that to make further progress it would be necessary to put a measuring electrode inside an axon, and the common squid, which has giant axons in its mantle with diameters of about half a millimeter, provided the opportunity. Hodgkin was now on the staff at Cambridge University and was joined by Andrew Huxley, his first student. They made some initial progress before the outbreak of war put the project on hold while they both worked, although in separate venues, on the development of radar.

The end of World War II brought them both back to Cambridge. Although work on the mechanism of the action potential in the squid axon in the United States had not been interrupted by the war, the problem was still waiting to be solved. Chapter 5, "The Mystery of Nerve Conduction Explained," describes how Hodgkin and Huxley used a technique known as the voltage clamp in experiments where wire electrodes were inserted into squid giant axons to reveal the universal mechanism of conduction of the action potential. Throughout this discussion the emphasis is on the properties of the action potential and what action potentials really look like as they travel along the axon. Why large-diameter axons such as the squid giant axon conduct action potentials faster than small-diameter axons and how the fatty insulation provided by myelin sheaths increases the conduction velocity of relatively fine vertebrate axons are explained.

Controversies in science often appear in either-or form. That was true for the question of how neurons communicate with each other. It boiled down to the issue of whether neurons are connected together in a continuous net allowing action potentials to be conducted directly between one neuron and the next or, alternatively, are separate individuals that might communicate with one another in a different way. Chapter 6, "Heart to Heart," begins with the work of Santiago Ramón y Cajal (1852–1934), possibly the most famous neuroscientist of all. (Ramón was his father's surname, which in some of his writings he has dropped in favor of just Cajal, his mother's.) Cajal was equipped with just a light microscope and a single technique, called the Golgi method, for silver staining individual neurons in dead specimens. Cajal was a genius of interpretation; he knew just what the appearances of the neurons were telling him. Among his many insights, Cajal figured out that neurons are individual entities, that axons are outgrowths from the cell bodies, and that the direction of travel of action potentials in most neurons is from the cell bodies toward the axon terminals. The story of Cajal segues into a modern description of the internal structure of the axon that lays the groundwork for the discussion of synaptic transmission.

Because neurons look like separate entities, interest arose in the question of how they might communicate with one another. Again there were two alternatives: some form of electrical transmission or possibly the release of a chemical from one neuron onto the next. The answer was found in a single experiment performed by Otto Loewi (1873–1961) on frog hearts—of all things! Frog hearts beat for a long time after removal from the frog to a simple salt solution. Although hearts beat without assistance from any nerves, two nerves innervate the heart: one that slows the heart rate and one that accelerates the heart rate. In an experiment that came to him in a dream Loewi showed that slowing the beat rate of an isolated heart for a period of time by electrically stimulating the vagus nerve released a substance into the salt solution that slowed the rate of a second heart that was beating without nerve stimulation. The substance turned out to be acetylcholine, which is also the neurotransmitter at the synapses between nerve and skeletal muscle and one of the major neurotransmitters in the central nervous system.

The detailed mechanism of synaptic transmission was elucidated in experiments with frog motor axons connected to muscle, that is, the frog neuromuscular junction. Just as in the case of the action potential mechanism, a way had to be found to measure the membrane potential with an electrode inside of a muscle fiber. Chapter 7, "Nerve to Muscle," begins with a description of the development of glass microelectrodes filled with a salt solution that could be inserted across the cell membrane. Starting with the experiments of Bernard Katz and Paul Fatt in the 1950s, a detailed picture of synaptic transmission at the neuromuscular junction was developed in which the arrival of the action potential at the axon terminal triggers the release of acetylcholine into the tiny gap between the terminal and the muscle fiber. The acetylcholine diffuses across the gap and binds to and opens receptor channels in the membrane that allow sodium to enter the muscle fiber, generating an action potential. The development of this scenario in its full detail demonstrates how this process can happen so fast that action potentials arriving at hundreds per second sum together to produce a strong contraction of the muscle. The experiments involved many feats of conceptual insight and technical genius, some of which are described. The chapter ends with a description of synapses in the central nervous system where individual synapses release far less neurotransmitter than at neuromuscular junctions. It takes the action of a great many synapses firing together to produce an action potential in the receiving neuron. There are also inhibitory synapses whose influence opposes the firing of an action potential by the receiving neuron.

Synapses in the brain and spinal cord in all their complexity are the substrate for modifications of the activity and connections of neurons that we call plasticity and learning. One of the major themes of twenty-first-century neuroscience is that the electrical activity of neurons changes the connections in the brain. Chapter 8, "Use It or Lose It," sets the stage for this story by describing some of the major modifications of the nervous system that occur during embryonic development. Next, the story moves to the experiments of David Hubel, Torsten Wiesel, and Simon LeVay on the visual cortex. Modification of the connections in the cortex was revealed by experiments where one eye of a cat or monkey was deprived of patterned vision for a period during postnatal development. The area innervated by circuits from the deprived eye was decreased because the innervation territory was taken over by circuits from the nondeprived eye. Functional magnetic resonance imaging studies have shown that such modifications occur in people with macular degeneration, in which the photoreceptors in the center of the retina degenerate. Axons originating from circuits connected to nondegenerated regions of the retina take over in regions of the brain that had been connected to the degenerated region of the retina. Chapter 8 ends with a discussion of the changes at synapses that occur during learning. Pioneering studies conducted by Eric Kandel and his colleagues have described learning in the large sea slug *Aplysia,* which takes place at individual synapses where two sensory inputs converge. If the activation of the first input is followed closely by the activation of the second input, mechanisms are triggered that produce a long-term increase in the amount of neurotransmitter that is subsequently released when the first input is activated. Thus, transmission through the synapses from the first input have been strengthened. The mechanism of synapse strengthening in the mammalian cortex during learning has also been investigated and involves both increases in the amount of neurotransmitter released and increases in the number of neurotransmitter receptors available to bind the neurotransmitter. Interestingly, the increase in the release of neurotransmitter during learning occurs in part because of the growth of new axon terminals—learning grows new connections in the brain.

All electrically excitable cells such as neurons and muscle fibers produce currents that flow through the extracellular environment during their activity. Chapter 9, "Broadcasting in the Volume Conductor," explores the significance and the uses of these broadcasts. It starts with two of the most extreme examples: the electric eels of the rivers of South America and the torpedo rays in the ocean worldwide. Torpedo rays have an interesting history because they were known to Europeans at least since ancient times and played a role in the early investigations of electricity. They produce strong shocks that deliver electric current into the water that they use to stun their prey. The wings of the ray contain massive electric organs, which are derived from muscle tissue. They produce high voltages by producing synaptic potentials on the ventral surfaces of disk-shaped cells called electrocytes that are stacked like poker chips spanning the thickness of the wings. When the nerve from the electric lobe of the brain fires the ventral synapses, the depolarization of the ventral surfaces turns the electrocytes into little batteries connected in series so that the voltages in a stack add together like batteries lined up in a flashlight, resulting in about +60 volts at the dorsal surface. Electric eels have longer stacks of electrocytes extending nearly the full length of their bodies, which can produce upward of 600 volts. Stunning shocks are not the only examples of animal electricity in the sea. With every heartbeat and every time a fish uses muscle to pump water through its gills, much smaller electric currents are broadcast into the surrounding water. A flat fish buried in the sand might seem hidden from predators, but sharks have electroreceptors on their heads that detect the tiny broadcasts, allowing them to dive into the sand and capture the hiding fish.

As far as anyone knows, the broadcasts from nerves and muscle inside our bodies have no function of their own—they are just consequences of the action potentials and synaptic potentials produced continuously within our bodies. However, they have important medical uses. The strongest broadcast is from the heart, which reaches the surface of the body and is detected by the electrocardiogram (ECG), which can be used to detect damage to the heart muscle. The electrical mechanisms that produce the heartbeat and the history of the development of the electrocardiogram are described. The other major broadcaster in the body is the brain, whose broadcasts through the skull appear on the surface of the scalp where they can be recorded by the electroencephalogram (EEG). The EEG has medical applications, but it does not reveal very much about the function of the brain. Chapter 9 ends with a discussion of some of the fringe areas that have developed from the study of electric fields within the body. While electric fields clearly can affect the function of cells, there is little evidence of this occurring in nature.

Chapter 10, "The Bionic Century," concludes the book with a discussion of the electronic restoration of functions of the nervous system that have been lost or severely compromised by injury or disease. The chapter begins with a discussion of the immense challenge presented by any regenerative approach to the restoration of function after spinal cord injury. The most viable approaches to improve independence and quality of life for spinal-injured individuals in the near future are likely to involve electronic prosthetic circuitry. Recently there has been tremendous interest in external circuits that detect the will to make movements in the electrical activity of the motor cortex of the brain and that produce that movement of the subject's own paralyzed limbs by activating stimulating electrodes implanted in the muscles. The pathway of the signal is brain to computer to muscles. Thus, the circuitry bypasses the damaged spinal cord and the peripheral nerves.

Experiments are described in which an array of recording electrodes was implanted in the motor cortex of a monkey and connected through a computer to stimulating electrodes implanted in the muscles that control grasping by the fingers. The monkey did not have a paralyzing injury. Instead, the fingers could be reversibly disconnected from the brain by injection of a local anesthetic into the motor nerves that innervate the muscles that move the fingers. Before injecting the local anesthetic, the investigators recorded the pattern of electrical activity in the motor cortex when the monkey performed a task that required picking up a ball and releasing it into a tube using his hand normally, without the nerve block. Then they set up computer algorithms to identify the electrical signatures in the cortex that accompanied the monkey's intention to pick up and to release the ball and used their appearance in the cortex to trigger the contraction and relaxation of the appropriate muscles. In effect, their goal was to read the monkey's mind and produce the appropriate finger movements using the connection through the external prosthetic circuitry, totally bypassing the spinal cord and peripheral nerves. When the fingers were disconnected from their motor innervation by the nerve block, the external circuit restored the monkey's ability to grasp the ball and release it into the tube. When the prosthetic circuit was disconnected, the paralysis returned. There are many variations of this kind of mind-reading prosthesis. Some involve effectors other than the subject's paralyzed limb, such as moving a robotic arm or a cursor on a computer screen. Mind reading was perhaps necessary to restore function in the monkey. However, paralyzed humans can learn to modify their brain activity to produce a desired movement.

Lastly, the chapter turns to bionic sensation. The most successful bionic replacement of any kind is the cochlear implant that restores a level of hearing that is good enough for speech recognition. How cochlear implants work is discussed in detail. The reason cochlear implants are so successful is that their input is electrical stimulation of the endings of the auditory nerve, and the sound signals are not processed until they enter the brain. The implant replaces the hair cells in the cochlea, which are the sensory cells that detect the sound. Hair cells are distributed in a spiral line along the basilar membrane such that hair cells that detect high-frequency sounds are at one end of the cochlea, hair cells that detect low-frequency sounds are at the other end, and there is a smooth progression of frequency response in between. The terminals of the axons of the auditory nerve fan out to innervate the hair cells, and axons (think of piano keys) at one end of the cochlea are stimulated to fire by high-frequency sound, axons at the other end are stimulated to fire by low-frequency sound, and there is a smooth progression of the frequencies of sound that cause firing of the axons in between. In cases of deafness where the hair cells have degenerated, a linear electrode array inserted into the cochlea, and extending most of its length, is used to stimulate the array of cochlear axons. An external microphone, similar to the microphone contained within an ordinary hearing aid, detects the sound. The microphone is connected to an external transmitter, which broadcasts signals through the skull to an internally implanted electrical stimulator. In response, the stimulator supplies electric current to the electrode array in the cochlea, distributing the stimulation among the array of axon endings (like fingers playing the piano) and mimicking the pattern of stimulation that had been originally provided by the hair cells. There are about twenty electrodes in the array, so the stimulation of the auditory axons is not as finely grained as that produced by the hair cells. The sound is different from normal hearing but good enough for speech recognition.

Vision is a far more challenging target for replacement by a sensory prosthesis. A retinal implant that replaces photoreceptors that have degenerated in diseases, such as retinitis pigmentosa, is a more difficult goal than the cochlear implant for two reasons. Vision has two dimensions instead of just one, and the retina contains neural circuitry that processes a 100-megapixel picture projected onto the retina into a 1-megapixel signal that leaves the eye via the axons in the optic nerve. Retinal implants under development employ a grid of electrodes representing far fewer pixels to directly stimulate the ganglion cells in the retina whose axons form the optic nerve. The picture is relatively coarse, and the processing of the retinal neurons is absent.

There are two types of retinal implant. In both, a grid of stimulating electrodes is implanted in the eye overlaying the retina. In one type, a stimulator that supplies current to the electrode array is controlled by the images captured by an external camera mounted on eyeglass frames. The other approach is the use of internal implants consisting of an array of stimulating electrodes overlaying the retina, with photodiodes in an array layered on top, which produce electric current in response to light and thus function as prosthetic photoreceptors. When light comes through the lens of the eye and strikes the photodiodes in a particular location, current is delivered to the electrodes at that location that stimulates the ganglion cells to fire, which sends action potentials to the brain signaling the detection of light at that location. In this way a pattern of light and dark projected on the prosthetic retina sends a crude representation of the pattern to the brain.

Both approaches have their advantages. Employing a camera to capture the image may provide more opportunity to computer process the image in ways that partially make up for the lack of processing by the original retina. The advantage of an implanted array of photodiodes is that a camera does not have to be pointed and focused since the eye itself provides these functions.

So far the goal has been to produce crude light and dark images of objects, but "vision" good enough to read large print may ultimately be possible, which would be a tremendous accomplishment. Efforts are also being made to input visual stimuli with surface electrodes on the visual cortex. This will likely be difficult because the cortical images are highly processed vector images rather than arrays of pixels. However, the exploration of possible interfaces between our bodies and the machines we manufacture is just beginning, and it is not possible to predict where the limitations lie. I do not think it is overly optimistic to believe that great advances that will improve the lives of those with disabilities will be forthcoming in the not-too-distant future.

1

ANIMAL ELECTRICITY

Anthropologists have never discovered a society where the people believe that they have no idea what the hell is happening. There are always explanations for mysterious phenomena handed down from past "authorities," and before the advent of the scientific method such explanations lasted for millennia. The function of the nervous system and muscle was "explained" by the animal spirits theory promoted by Praxagoras (b. 340 B.C.E.), a younger contemporary of Aristotle (384–322 B.C.E.), and by Erasistratus (304–250 B.C.E.), Aristotle's son-in-law (Smith et al. 2012). Erasistratus dissected executed criminals and came to believe that the body is made of tubes and all mechanisms are basically plumbing. This conclusion was not unwarranted given all the fluid and tubes that he encountered during his dissections and the mechanisms he had at his disposal. Apparently it was not easy to understand what filled some of the tubes, even in the case of the most obvious body fluid—blood. Praxagoras recognized the difference between arteries and veins, but the arteries he encountered during his dissections were empty, likely because the dissection technique used in the Lyceum produced arterial spasm, causing the arteries to collapse and forcing the blood they had contained to engorge the veins. Thus, the veins could well have appeared to be the major carriers of blood, while the arteries with their thicker walls must have held the evanescent pneuma, the breath of life, collected by the lungs but now escaped into thin air through the incision made during the dissection. You might ask: Why does blood spurt out when an artery is cut in a living animal? Because, Praxagoras believed, the escaping pneuma is replaced by blood extremely fast in the living animal, as if it was never there. Erasistratus went on to identify three fluids: the most obvious but lowly blood, which is the nutritive fluid produced from food that keeps the body alive; *pneuma zootikon,* the vital spirits, captured by the lungs (oxygen was not discovered until the latter part of the eighteenth century); and *pneuma psychikon,* or animal spirits, produced from vital spirits by the brain, which is responsible for consciousness, thinking, sensation, and movement. *Animal* was not used in the zoological sense. *Anima* is the Latin translation of the Greek *psyche.* The psyche is the human mind, and animal spirits are the spirits that animate us.

No one is more responsible for perpetuating the view of the nervous system as plumbing carrying animal spirits than Claudius Galen (129–ca. 216 C.E.), one of the fathers of medicine. Galen was born into a wealthy family in the ancient Greek city of Pergamon, a cultural and intellectual center. His rich, ambitious, architect father chose medicine for Galen's career. Galen cooperated, graduated from the medical school at Alexandria, and then returned to Pergamon for five years where he served as a physician to gladiators, so he knew something about blood and guts. Then he moved to Rome where he became a physician to a series of emperors and possibly the most prolific writer of medical books that ever lived. His writings propelled Galen's influence on medical thinking for over a millennium, until the Renaissance.

It is customary to credit Francis Bacon (1578–1626) with originating the scientific method during the Renaissance, but experimentation has likely always been with us. Galen himself was something of an experimentalist and made some advances. He discovered that intact arteries contain blood by tying off an artery in two positions in a living animal and showing that the section of artery in between bled when punctured. Thus, the section of artery must have contained blood when it was tied off, before the blood had a chance to replace the escaping vital spirits. Galen also discovered some of the circuitry of the nervous system by showing that certain parts of the animal body were paralyzed when the nerves supplying them were cut. He showed conclusively that the brain, not the heart, is the recipient of sensation and the source of willful movement, contrary to the beliefs of Aristotle and many others. However, Galen perpetuated the belief that motor commands were carried by animal spirits flowing to the muscles through hollow nerves. Discoveries in the fields of electricity and cell biology that would eventually provide the framework for explaining the functioning of nerve and muscle would not begin to accumulate until the eighteenth century. The actual electrical mechanism of conduction in nerve was not established until the mid-twentieth century. In the meantime, many more physicians, philosophers, and other thinkers threw their ideas into the ring. What emerged was a slowly evolving, often highly nuanced, but always fundamentally wrong picture of how nerves and muscles work.

As the ideas of Praxagoras, Erasistratus, and Galen illustrate, in our attempts to explain the world, none of us has any option other than to rely on the mechanisms that we know about, or believe we know about. The ancients knew about the world available to the unaided senses. When they evoked unseen entities to explain how the body and brain work, those entities were simply unseen forms of what they were familiar with. Thus, nerves were considered to be tubes through which animal spirits flowed like water through a pipe. One might fault the ancients for believing in ethereal entities they could never observe that travel through tubes in the body, but no one has ever seen an electron, and the laws of electricity were established long before anyone knew what electric charge flowing through wires really is. The difference lies in the framework through which people see the world and the methods they deem appropriate to build and refine that framework. Until the seventeenth century the results of experiments were not given much credence against the word handed down from ancient authorities.

The unseen world of the scientific era is replete with entities that are unavailable to the unaided senses. A huge part of science has been occupied with making them visible in the broadest sense. In some cases they became literally visible as, for example, the cell and some of its internal structures that were seen under the microscope. Structures a little bit smaller such as the cell membrane are beyond the resolution of the light microscope and only become "visible" in pictures of dead cells taken with the electron microscope. Objects on the molecular, atomic, and subatomic scales have been made "visible" by their influence upon the visible world; for example, individual electrons were never "seen" until their presence was made "visible" in the Millikan oil-drop experiment (see Chapter 2).

In order to explain how electricity makes our bodies and brains work we must delve into the world of electricity and the worlds of the cell and the molecules and atoms that make cells work the way they do. These worlds have rules of their own that are in many respects unlike the rules governing the world available to our unaided senses. We begin with the world of electricity. It is one thing to "understand" electricity by memorizing a few rules about how electric current flows in circuits, but it is quite another to develop the imagination required to understand what electric current would look like if we could somehow get down to its level and actually see it. Although it will be necessary for readers to understand the workings of a simple circuit diagram, it is the view of electricity that is only available to the imagination that we are really after. This view, along with the view of cells, molecules, and atoms that is developed in Chapter 2, will set the stage for describing how nerve cells make and conduct the electricity to produce our sensations and actions and everything that goes on in between.

THE HEAT THEORY OF THE ELEVATOR BUTTON

I began my personal inquiries into the foreboding world of electricity at UCLA where I was a graduate student in physiology in 1968. One day a fellow student and I began some tests of the heat theory of the elevator button. We were not the first to develop this theory; it was surely developed independently by many people. Although this theory is as wrong as the animal spirits theory of nerve conduction, I occasionally run across someone who still believes in it today. The source of the great mystery was the elevator buttons that light up when touched without any detectable mechanical movement. How do they sense the presence of a finger? Like Praxagoras, Erasistratus, and Galen, we took the only approach that humans can when trying to figure out a mystery—we tried to explain it in terms of phenomena already known to us. This was long before the era of personal computers with track pads and touch screens. Now the phenomenon is so commonplace, and therefore accepted, that nobody bothers to think about how nonmechanical buttons work, and I am sure the vast majority of computer users have no clue. Back then the nonmechanical elevator button was just about the only example, so familiarity had not yet obscured its mystery.

My friend and I hypothesized that nonmechanical elevator buttons work by sensing the warm temperature of a finger. Like other primitive people of the time, we focused on heat rather than electricity because the plastic surface of the button could certainly conduct heat. We couldn't see how it could conduct an electrical signal; since plastic is an electrical insulator we "knew" that electric current cannot pass through it. Moreover, we could not see how our finger could produce an electrical signal when we touched the elevator button since we were not connected in a circuit, and we "knew" that a person has to be "grounded" to get a shock. As I remember, the experimental basis for our belief in the heat theory was primarily the failure of the hard press with the finger of a gloved hand to light up the button, but the data were confusing because occasionally even a bare finger failed to light up the button. In some cases this could be explained because the finger, having just come in from the outdoors, may have been too cold. But other times it seemed capricious whether the button would light up or not when touched by a bare finger. It never occurred to us that the electrical insulation provided by dry skin could be the problem. We eventually forgot about the issue and moved on to more productive endeavors.

My greater familiarity with the principles of electricity along with the prevalence of track pads and touch screens have caused me to reconsider the heat theory of the elevator button. The fine discriminations made by track pads and touch screens must certainly rule out finger temperature as their operating principle. My new theory, again not original with me, is the capacitance theory of the elevator button.

Unlike investigations of natural phenomena, where the truth of an explanation must rest solely on observation and experimental evidence, elevator buttons are human inventions, so we can read about how they really work, and the Web makes explanations of such things readily accessible. Back in the 1960s it was just too much trouble for me to find out, but now I know that the capacitance theory is correct. The concept of electrical capacitance explains how electric current can pass through a thin insulating material like the plastic surface of the button and how contact with our bodies, even when we are not grounded, can alter the behavior of a circuit behind the surface of the button, signaling that someone wants to ride the elevator.

VOLTAGE, CHARGE, CURRENT, RESISTANCE, AND CAPACITANCE

Nonmechanical elevator buttons have several features in common with the conduction of impulses along nerves and muscle fibers. Most important, both involve the flow of electricity through body fluids and both involve the charging and discharging of a capacitor. Elevator buttons are simpler than nerve and muscle, so let's begin by dispelling the great mystery behind elevator buttons that make no movement when pressed. A few concepts of electricity are necessary: namely, voltage, charge, current, resistance, and capacitance. Understanding capacitance requires understanding the other four, so we will deal with capacitance last.

German physicist Georg Ohm (1789–1854) first published his law in 1827. Ohm's law is one of those delightfully commonsense relationships of the following form: the flow of something is determined by the combined effect of the force driving the flow and the resistance opposing the flow. The familiar form of Ohm's law is $I = V/R$, where electric current *(I)* is equal to the voltage *(V)* driving the current divided by the resistance *(R)* to the flow of current (Figure 1.1). This makes sense; current is doubled by doubling the voltage driving it and halved by doubling the resistance against it. The same kind of relationship holds for water flowing through a pipe where the amount of water flowing is determined by the water pressure driving the flow divided by the resistance in the pipe opposing the flow.

$$I = \frac{V}{R}$$

$I =$ current (amperes)

$V =$ voltage (volts)

$R =$ resistance (ohms)

Figure 1.1. Ohm's law.

Ohm's law seems simple and obvious enough to us now. However, it was not immediately accepted, and Ohm was ridiculed as unfit to teach science because he was promoting "a web of naked fancies [and] an incurable delusion, whose sole effort is to detract from the dignity of nature" (Davies 1980, 57). Heaping invective in print upon another scientist was more or less normal scientific discourse at the time.

Electrons were discovered in 1897 by a team of British physicists led by J. J. Thomson. When Ohm developed his law he did not know what was actually flowing in wires, but he did have the means to apply a voltage and indirectly measure the resulting current flow in a circuit. When Ohm measured the current driven through a wire by a certain voltage, the current doubled when he doubled the voltage, and when he doubled the length of wire, the current was reduced in half, so the resistance must have doubled. Ohm also found that the same lengths of wire made of different metals had different resistances; some metals are better conductors of electricity than others.

Everything in our world that we can touch is made of atoms, and atoms consist of nuclei made of protons and neutrons surrounded by orbiting electrons. We now know that all negative charge is in the form of electrons, all positive charge is in the form of protons, and the charge of a single electron equals the charge of a single proton. It is a fundamental rule of physics that like charges repel each other and unlike charges attract each other; that is, electrons repel each other, protons repel each other, but electrons and protons attract each other. I will bet you have no idea of the immense strength of electrical repulsion and attraction. Their true magnitude came as a shock to me when I first encountered the following quotation from *The Feynman Lectures on Physics* (Feynman, Leighton, and Sands 1964, 1.1):

Consider a force like gravitation which varies predominantly inversely as the square of the distance, but which is about a *billion-billion-billion-billion* times stronger. And with another difference. There are two kinds of "matter," which we can call positive and negative. Like kinds repel and unlike kinds attract—unlike gravity where there is only attraction. . . .

There is such a force: the electrical force. And all matter is a mixture of positive protons and negative electrons which are attracting and repelling with this great force. So perfect is the balance, however, that when you stand near someone else you don't feel any force at all. If there were even a little bit of unbalance you would know it. If you were standing at arm's length from someone and each of you had one percent more electrons than protons, the repelling force would be incredible. How great? Enough to lift the Empire State Building? No! To lift Mount Everest? No! The repulsion would be enough to lift a "weight" equal to that of the entire earth!

Electrons are the only charged particles that are free to move in a solid conductor such as a metal wire. The movable electrons originate from the outer electron orbitals of atoms that make up the solid structure of the metal. In contrast, the atomic nuclei containing the protons are immobilized in the structure of the metal, so positive charge cannot move along a wire. Think of a wire as a solid matrix of fixed positive charges, which are neutralized by an equal number of mobile electrons, so the wire has no net electric charge. The electrons are free to move around, exchanging positions with one another, but on average they are spread evenly throughout the wire. If the wire is used to connect the negative and positive terminals of a battery, electrons will flow from the negative terminal into the wire, and an equal number of electrons will flow from the wire into the positive terminal. During the flow of current, there will be only a very slight increase in the number of electrons in the wire over the number of fixed positive charges. As electrons move through the wire, every time an electron leaves the vicinity of a fixed positive charge, another electron comes along into the position vacated by the departing electron, so the fixed positive charge remains neutralized by a nearby electron at all times. If this were not the case, the departing electron would be held back by its strong attraction to the fixed positive charge and no current would flow. The metaphor that compares electrons flowing through a wire to water flowing through a pipe breaks down when you disconnect or cut the wire and the electrons do not spill out because the immense electrical attraction of the protons in the structure of the metal holds them tightly inside.

For Ohm's law (Figure 1.1) to work, current, voltage, and resistance all need units of measure. Since no one knew what was flowing when the units were developed, the unit sizes that were picked are somewhat arbitrary. The scheme that was developed has a voltage of one volt producing a current of one ampere when the resistance of the circuit is one ohm. Since current is the flow of charge, the ampere needs further definition: one ampere of current is the flow of one coulomb of electric charge per second. The relationships of these units to one another are relative, and they need an absolute reference to set their scale. The unit of mechanical power used in physics, the watt, sets the scale of the electrical units. One watt of power is produced when one ampere of current flows across a voltage difference of one volt. Scaling the electrical units to the watt produces numbers of manageable size.

A few years after the discovery of the electron, Robert Millikan (1868–1953) performed experiments in which he measured the electric charge in coulombs of a single electron (see Chapter 2). It turned out that a coulomb of charge is equal to the charge of approximately 6.24 quintillion (6,240,000,000,000,000,000) electrons. One ampere of current is the flow of 6.24 quintillion electrons every second! The typical household circuit puts out 15 times that. You can see why the units of electricity are not scaled to the charge on a single electron.

The sources of voltage in the circuits that we will be dealing with in nerves and muscle behave like batteries and have plus and minus terminals just like the AA batteries you might buy to power a flashlight (Figure 1.2). Benjamin Franklin (1706–1790) devised the plus and minus designation of electricity (Smith et al. 2012). Franklin believed that electricity flowed from objects that have an excess of electric matter, which he designated positive, into objects with a deficit of electric matter, designated as negative. This made sense at the time—positive objects contained something that is absent from negative objects. Franklin's

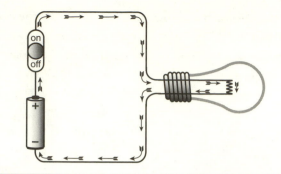

Figure 1.2. Circuit connecting a battery through a switch to a light bulb.

designation was retained after the discovery that negative charges in the form of electrons flow in wires. Paradoxically, electric current in wires is still conceptualized as positive charge flowing in the opposite direction of the actual flow of electrons. When the switch in Figure 1.2 is turned on, current flows clockwise around the circuit (arrows) from the positive terminal of the battery to the negative terminal. This makes no difference in the analysis of the flow of electric current since it turns out that the effects of positive charge flowing in one direction are equivalent to the effects of negative charge flowing in the opposite direction. Confusing? Yes, but there is a silver lining for readers of this book. By chance, this convention makes it a little easier to talk about animal electricity since, as we shall see, the currents carried by the positively charged ions, potassium and sodium, play major roles in generating electrical events in animals, and the direction of current is the same as the actual flow of these ions. We will get to ions below.

The voltage specified in Ohm's law refers to a difference in voltage between two locations, which, in the case of a battery, are the positive and negative terminals. The chemical reactions inside a 12-volt battery, such as the one in your car, produce a difference of 12 volts between the terminals. The terminal marked with a plus is 12 volts positive relative to the terminal marked with a minus. This 12-volt difference is entirely relative. You could consider the negative terminal as having a voltage of zero and the positive terminal as having a voltage of +12 volts, or you could consider the positive terminal as having a voltage of zero and the negative terminal as having a voltage of −12 volts. You could even consider the negative terminal to have a voltage of +100 volts and the posi-

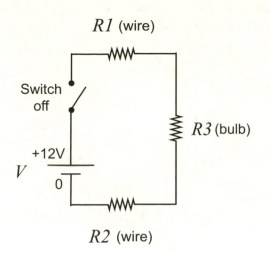

R1 (wire)

Switch
off

+12V

V

0

R3 (bulb)

R2 (wire)

Figure 1.3. Circuit diagram representing the light bulb circuit with the switch off.

tive terminal to have a voltage of +112 volts, or the negative terminal to have a voltage of −112 volts and the positive terminal to have a voltage of −100 volts. As long as the voltage of the positive terminal is 12 volts greater than the voltage of the negative terminal, the current flow in the circuit will be the same. Note that whichever of these combinations of voltages are used, the positive terminal is always called the positive terminal and the negative terminal is always called the negative terminal because in every combination the positive terminal is positively charged relative to the negative terminal.

Now we can construct a diagram of a circuit with a 12-volt battery and a series of three resistors (Figure 1.3). The battery symbol is a pair of parallel lines, the longer line indicating the positive terminal and the shorter line indicating the negative terminal. Zigzag lines indicate resistors. The voltage of the negative terminal is set at zero, and the positive terminal has a voltage of +12 volts.

Imagine that resistor R3 in Figure 1.3 represents the resistance of a light bulb—maybe the headlight of a car. The lines along which current flows in circuit diagrams do not represent wires because wires have resistance. Lines just indicate electrical continuity, and the voltage along a continuous line is always the same. Let's designate resistors R1 and R2 as representing the resistances of the wires connecting the light bulb to the battery. The resistance of a light bulb filament is so much greater

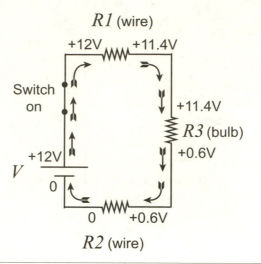

Figure 1.4. Circuit diagram representing the light bulb circuit with the switch on.

than the resistance of a suitable length of wire, that the resistance of the wires is usually insignificant and could have been left out of the circuit diagram, but for our purposes we need to assign the wires a significant resistance. Let's assign the wires as each having 5% of the resistance of the light bulb filament. When the switch is in the off position, there are no voltages across the three resistances.

When the switch is turned on (Figure 1.4), current flows clockwise around the circuit as indicated by the arrows. It is a principle of electricity that all of the current that departs from the positive terminal of the battery must arrive at the negative terminal of the battery; therefore, an identical amount of current must flow through each resistor. Since the current through each resistor is the same, Ohm's law tells us that the voltage difference across each resistor, divided by the value of its resistance, must be the same for R1, R2, and R3. Since R1 represents 5% of the total resistance in the circuit, then the voltage drop across R1 must be 5% of −12 volts, which is −0.6 volts. The voltage drop across R3 must be 90% of −12 volts, which is −10.8 volts. Finally, the voltage drop across R2 is −0.6 volts, the same as the voltage drop across R1, which brings the voltage to zero once the current has passed through all three resistors. As shown in Figure 1.4, measurements of the voltage around the circuit while the current is flowing would show +12 volts at the positive terminal of the battery, +11.4 volts after crossing R1, then +0.6 volts after crossing *R3,* and finally 0 volts after crossing *R2.* Two important principles emerge from all this: The first principle is that the voltage drops around the circuit have a total value equal and opposite to the voltage of the battery driving the current; that is, the voltage in this circuit must be zero after all the resistances in the circuit have been crossed by the current. The second is that the share of the total voltage drop across each resistor is proportional to each resistor's share of the total resistance, which is a simple consequence of Ohm's law.

Now *imagine* you are standing over a 12-volt battery that has been disconnected from your car and you simultaneously touch one of your hands to the positive terminal and the other to the negative terminal *(Do not try this!).* You have now completed the circuit and your body has become the only resistor. The 12 volts will drive current through your body that you will definitely feel as a shock. I can vouch for this because I once accidentally dropped a Leatherman plier tool across the terminals of a car battery. This produced a tremendous spark while burning a one-cubic-millimeter notch in the steel handle *(Do not try this either!).*

Now if you only touch the positive terminal of the battery with one hand, and do not touch the negative terminal, you will not complete a circuit between the positive and negative terminals. You will feel nothing, and you may conclude that no charge flows between the battery and your body because you don't feel anything and because you "know" that charge can only flow if there is an uninterrupted connection from one terminal of the battery to the other. This is why a light bulb connected to one of the terminals of a battery in a flashlight does not shine until you turn on the switch connecting the light bulb to the other terminal of the battery. But it is incorrect to conclude that charge does not flow into your body when you connect yourself to only one terminal of the battery. If you touch only the positive terminal, enough current will flow to charge your body to the same voltage as the positive terminal. Then the flow of current will stop because current only flows between two locations in a circuit that have different voltages. Now, if you let go and do not touch anything that can discharge your body, your body will remain charged to the same voltage as the positive terminal. If you then touch only the negative terminal, which is 12 volts negative relative to the charge now on your body, current will flow until your body is charged to the same voltage as the negative terminal. Since you did not touch both terminals simultaneously to establish a continuous flow, the flow of current in this scenario would be extremely small—much too small to feel. The amount of electric charge in coulombs needed to charge your body to 1 volt is the capacitance of your body in farads. The capacitance represents the capacity of your body to store charge.

THE CAPACITANCE THEORY OF THE ELEVATOR BUTTON

Now back to the elevator button. A metal plate lies behind the thin, plastic surface of the elevator button. Like all conductors, the plate has capacitance so it will contain a certain amount of electric charge depending upon its voltage. An electric circuit behind the button connected to the plate supplies the charge. Although the electrons that flow through wires into and out of the metal plate cannot pass through the plastic surface of the button, the plastic surface is only a fraction of a millimeter thick, so a finger placed on the surface of the button is close enough to the plate to increase the plate's capacity to hold charge at a given voltage.

This is how it works. Your finger is a conductor of electricity because it contains electric charges that are free to move, but the flow of electricity in your finger is nothing like the flow of electrons along a metal wire. Much of the volume of your finger is fluid containing dissolved ions. The fluids are the cytoplasm inside the cells, the extracellular fluid surrounding them, and the blood. Ions are atoms or molecules that have a net electric charge because the number of electrons and protons they contain do not match. A hydrogen ion (H^+) is a single proton free in solution. Other positive ions such as sodium ion (Na^+), potassium ion (K^+), and calcium ion (Ca^{+2}) are charged because they contain more protons than electrons—one more in the case of sodium ions and potassium ions and two more in the case of calcium ions. Unlike a proton, an electron by itself is not an ion, and electrons do not move through the body alone except in very special circumstances such as when you receive an electric shock. Under normal circumstances the movement of electrons in your body occurs as the movement of negatively charged ions such as chloride ion (Cl^-), which contains one more electron than protons. There are many other ions dissolved inside your cells and extracellular fluid that can move when you touch the elevator button, but sodium, potassium, calcium, and chloride all have especially important roles to play in animal electricity.

When electrons flow into the plate behind the thin plastic surface of the elevator button, the plate becomes negatively charged, and the electric field emanating from the negative charge repels negative ions, for example, chloride, in your fingertip, causing them to move through your finger away from the tip, and attracts positive ions, for example, sodium, in your finger, causing them to move toward the tip. Thus, the result of touching the button is that excess positive charge accumulates in your fingertip close to the button, partially neutralizing the negative charge on the plate. Because some of the charge on the plate is neutralized by the presence of a finger, more negative charge will have to flow into the plate to charge it to a given voltage; that is, the touch of your finger increases the capacitance of the plate.

What exactly is this electric field that extends through the thin, plastic surface of the button between the charged plate and your fingertip? In the classical Newtonian analysis, electric fields, magnetic fields, and gravitational fields are force fields where particles of matter exert forces on one another without touching, known as "action at a distance." The magnitude of force generated by a single electron equals the force generated by a single proton, and the distribution of the force around an electron or proton decreases in strength with the square of the distance between the charged particles. Thus, when the distance doubles the force drops fourfold. When an ion with an extra electron like chloride or an extra proton like sodium in the fingertip is moved a small distance away from the plate, the force exerted on the ion by the electric field emanating from the plate is greatly reduced. This is why the plastic surface of the button has to be thin.

The plate behind the surface of the elevator button is supplied with alternating current, so its voltage alternates between positive and negative. Although protons are not free to move through metal wires into the plate, the plate becomes positively charged when electrons are drawn away from it through the circuit, leaving behind a slight excess of positive charge in the form of the protons in the immobile atoms making up the structure of the metal. When the plate is positively charged, negatively charged ions will flow along the finger into the fingertip that is touching the plate, and positively charged ions will flow along the finger away from the fingertip—the opposite of the flow of charge in the finger that occurs when the plate is negatively charged. The current provided in ordinary circuits into which we plug electric lights and appliances is an alternating current that reverses in polarity 60 times a second (called 60 cycle). So it is with the plate behind the insulating surface of the elevator button, although the voltage is much lower than the 120 volts supplied at an ordinary electric socket. As the electric circuit behind the button drives the plate to swing back and forth from positively charged to negatively charged, the electric field projected across the plastic surface of the elevator button into the finger swings from positive to negative, and the ion movements in the finger swing back and forth as a result. Sodium will move toward the fingertip when the plate is negative and away from the fingertip when the plate is positive, while chloride will move away from the fingertip when the plate is negative and toward the fingertip when the plate is positive.

By now you are probably beginning to see how cumbersome it is to describe electricity in terms of the movement of charged particles, since the particles that move are different in different materials—for example, wires versus fingers. Fortunately, the concept of electric current comes to the rescue. Analyzing the elevator button and finger system in terms of current rather than the flow of electrons and ions, we would now say that when current flowing from the circuit behind the button into the plate causes the plate to become positively charged, current then flows along the finger away from the fingertip. Conversely, when current flowing from the plate into the circuit behind it causes the plate to become negatively charged, the current in the finger reverses, flowing along the finger into the fingertip toward the button.

We can see in this circumstance that it looks as if current is flowing through the plastic surface of the button, even though no charged particles can flow through this insulator. However, the plastic surface is so thin that the electric field can cross it, which itself can be viewed as a form of current. So in the conceptualization of electric current, there is an electric current flowing between the plate and the finger, even though no charged particles actually flow through the insulator separating them. The configuration of two conductors separated by a thin insulator forms a capacitor, and the current that flows through a capacitor is called capacitative current.

The presence of a finger touching the elevator button creates a capacitor, which affects the flow of current in the circuit connecting to the plate behind the button. This circuit is configured in what is known as a timing circuit. Current flowing through the capacitor formed when a finger touches the elevator button slightly alters the timing of the alternating current flowing through the timing circuit, which is the signal that activates the button causing it to light up and summon the elevator. Simply stated, the elevator button circuit is able to detect the small alternating capacitative current passing between the plate in the elevator button and your fingertip.

Two ideas are crucial to understanding the elevator button. One is the realization that electric current is a concept, not a physical entity in the ordinary sense, the way an electron or proton is a physical entity. Current can flow in wires in the form of moving electrons, in liquids in the form of moving positive and negative ions, and across a capacitor in the form of an electric field without the flow of electrons or ions. The other is that current can flow into or out of an object that is not connected in a complete circuit—the finger in our example. This analysis of the elevator button makes it clear that I would never have been able to figure out how nonmechanical elevator buttons work when my knowledge of the possibilities was limited to the hypothesis that the button sensed finger temperature. Only after I gained knowledge of the relatively sophisticated concept of capacitance did I have any chance.

Capacitors used in most electric circuits, of course, do not involve the use of fingers; they consist of two conductive plates separated by an insulator. Since electric charge cannot actually cross the thin insulator between the plates of a capacitor, current entering the capacitor causes positive charge to accumulate on one plate, while the current leaving causes positive charge to be depleted from the opposite plate. Thus, any time current is flowing through a capacitor the voltage difference between the plates is changing because of the accumulation of charge on one plate and depletion from the other plate. If the voltage across a capacitor is constant, that means that no charge is accumulating or dissipating from the plates, so no current is flowing across it. Cell membranes form the insulator between the cytoplasm inside of the cells and the extracellular fluid that bathes them, and current flow and voltage changes across the capacitance and resistance of cell membranes are the major expressions of animal electricity, as we shall see.

Capacitors have an interesting property when connected in a circuit to a source of voltage such as a battery. Figure 1.5 shows the circuit from Figures 1.3 and 1.4 with resistor *R3* replaced by a capacitor *(C)*. Not surprisingly, the symbol for a capacitor is two parallel lines representing two conductive plates across the current path with the space in between representing the insulator that prevents electrons or ions from crossing between them. Only capacitative current carried by an electric field can cross a capacitor. When the switch is in the off position, as in Figure 1.5a, no current is flowing. If the switch is turned on, current will briefly flow through the circuit in the clockwise direction, and the 12-volt drop in voltage will be apportioned across *R1,* the capacitor, and *R2.* At the instant the current begins to flow (b), no electric charge will have had time to accumulate across the plates of the capacitor, so the voltage drop across the capacitor starts out at zero, and the entire drop of 12 volts will be apportioned equally, 6 volts each, to *R1* and *R2* because their resistances are equal. However, this situation changes rapidly as the charge delivered by the current accumulates across the plates of the capacitor. When the capacitor is charged to 6 volts (c), there is a voltage drop of 3 volts driving current across each resistor, so the current flowing in the entire circuit is reduced by half. When the capacitor is charged to 12 volts (d), the current stops flowing, even though the switch is still turned on. Applying Ohm's law to what is happening during the time between turning on the switch and charging the capacitor to 12 volts, you can see that as charge accumulates on the plates and charges the capacitor there is a smaller and smaller voltage drop left over to be apportioned across the two resistors. With less voltage drop across the resistors, the current flowing through them becomes smaller and smaller, until the capacitor is charged to 12 volts, and there is no voltage left to drive current through the resistors, so the current stops flowing. Since there is no current flowing, it does not matter whether the switch is on or off, so you can turn the switch off and the capacitor remains charged at 12 volts. The end result is electric charge stored across the capacitor.

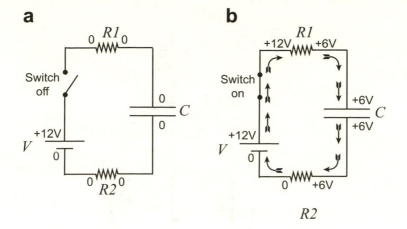

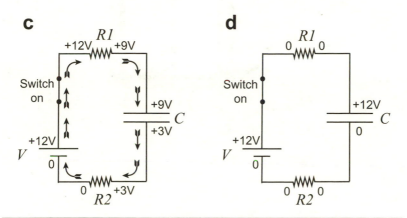

Figure 1.5. Circuit diagrams representing a battery connected through a switch to a capacitor.

You can disconnect the capacitor, take it away, and connect it into another circuit into which it will deliver its stored charge.

THE LEYDEN JAR

Take a good long look at the circuit in Figure 1.5 because it is the only circuit you will need to know to fully understand how the nervous system conducts electricity, but that is not until Chapter 3. First let's dial back to where capacitors came from. The capacitor was developed

Figure 1.6. Diagram of a Leyden jar.

long before the battery. The invention of the Leyden jar, independently in 1745 by Pieter van Musschenbroek (1692–1761) in the Netherlands (hence the name taken from the Dutch city) and Ewald Georg von Kleist (1700–1748) in Pomerania, made the study of electricity really take off. Before the development of the Leyden jar, electricity could only be obtained at its source, which was usually a friction machine that generated static electricity. In 1672 Otto von Guericke (1602–1686) published a description of the first electric generator, which consisted of a friction machine in which a large rotating sulfur ball rubbed against the hand (Brazier 1984). Many different generating machines were developed in the eighteenth century using a variety of materials, but all were mechanical devices that produced friction, which we now know causes electrons to rub off one material and collect on another.

As the name implies, the Leyden jar is a jar (Figure 1.6), but that was not essential. The Leyden jar is a capacitor consisting of a three-layer sandwich where two materials that conduct electricity are separated by an insulator. The glass wall of the Leyden jar served as the insulator.

Covering the outside surface of the jar with metal foil and lining the inside surface with metal or filling the jar with small metal balls produced the sandwich.

Leyden jars were electrically charged by connecting a static electric generator to terminals that were electrically connected to the metal on the outside and inside of the jar. When the generator was disconnected, the charge remained until a conductor of electricity such as a metal wire or the human body bridged the terminals. Quite high voltages could be established in Leyden jars. Dramatic demonstrations were in vogue. Abbé Jean-Antoine Nollet (1700–1770) electrified 200 monks, each connected to the next by a 25-foot length of iron wire. This produced a loop nearly a mile long. When the monks at each end were connected to the poles of a Leyden jar, every monk felt the shock.

The ability to produce and store electricity at will created the opportunity to experiment with electricity and discover many of its properties, but before that was accomplished electricity became a form of entertainment. A Scottish showman, calling himself Dr. Spencer, demonstrated sparks flying from an electrically charged boy suspended on a wire in midair. His show captured the imagination of an uneducated runaway from a dysfunctional family who was in a Boston audience in 1743. This was 20 years after Benjamin Franklin had run away, and he had established a successful printing business in Philadelphia and founded *Poor Richard's Almanack* and the *Pennsylvania Gazette*. He was so taken by the electrical demonstrations that he advertised Dr. Spencer's shows and lectures in his publications. Not long afterward, Franklin purchased all of Dr. Spencer's electrical apparatus and began a personal study of electricity. The accepted test for bona fide electricity was the production of a spark. Franklin became famous for his experiment where he showed lightning is electricity by flying a kite in a thunderstorm, which resulted in electricity being conducted down the wet kite string to electrify an attached key (Franklin 1753). As this example shows, the value of popularizing science should never be underestimated because it occasionally attracts the attention of people who go on to make major contributions to knowledge.

Many of those involved in the early days of electricity jumped to the conclusion that electricity could be used to treat medical conditions. Abbé Nollet, of electric monk fame, originally believed that electric shock might cure paralysis but eventually became skeptical. Benjamin Franklin himself started an electrotherapy program where he tried to cure patients with palsied or even paralyzed limbs, but he soon concluded that the benefits were only temporary. Even so, a huge cottage industry developed, populated by many clear or alleged charlatans. Among them was Franz Anton Mesmer (1734–1815), who invented his own version of animal electricity, which he called "animal magnetism," upon which he based treatments for conditions ranging from psychological problems to blindness. His name, of course, is the root of the term *mesmerize*. Mesmer caused such controversy in Paris that in 1784 Louis XVI commissioned Benjamin Franklin and the renowned chemist Antoine Lavoisier to investigate. Their approach was very modern. They tried and failed to detect animal magnetism with a range of instruments. Using double-blind experiments—highly innovative for the time—they were unable to demonstrate that animal magnetism had any effect on patients.

ANIMAL ELECTRICITY

In the eighteenth century nobody had any idea what electricity was, and its manifestations were so varied, from rubbing a cat's fur to lightning, that it was hard to believe that electricity could be a single entity. The only occurrences of animal electricity readily observable in the eighteenth century were the shocks produced by electric fish. Electric torpedo rays of the Mediterranean Sea and Atlantic Ocean were known in Europe since ancient times. Most species of torpedo ray generate relatively low-level shocks of about 60 volts, which produce strange feelings of cold and local anesthesia when a hand touches the fish or is placed nearby in the water. Galen thought that torpedo rays produced their effects by releasing poison, which seems consistent with the effects. After all, the modern term for injecting local anesthesia for dental work and minor surgery is *freezing*. Even though torpedo rays produce a higher voltage than a car battery, they deliver far less current and did not produce shocks powerful enough to generate a spark easily detectable with eighteenth-century methods. The dangerous, walloping shock of about 600 volts produced by electric eels would be more difficult to explain away by nonelectrical mechanisms. But electric eels live in the rivers of

South America and were hard to study because they tended to die during the voyage to England.

Eventually living eels were successfully transported to England, and in 1776 John Walsh, who worked under the mentorship of Benjamin Franklin, was able to demonstrate a spark from a Leyden jar that had been charged by an electric eel. Franklin himself, who spent much time in Britain and continental Europe, missed the demonstration because he was sailing back to America in order to be on his country's side of the Atlantic for the impending revolution. The demonstration of the reality of animal electricity in electric fishes raised the possibility that electricity, in more subtle form, may be involved in the functioning of nerve and muscle. In fact, in a prophetic statement back in 1671, Francesco Redi (1626–1697) of Florence, who is remembered mostly for his experiment disproving the spontaneous generation of maggots, described the torpedo electric organ as possibly being a type of muscle, which turned out to be correct (Smith et al. 2012). Thus it was becoming reasonable to imagine electricity playing a role in the normal functioning of nerve and muscle.

The stage was set for the appearance of Luigi Galvani (1737–1798). Compared to Franklin, Galvani had a more ordinary career development. He graduated from medical school in Bologna, his place of birth, worked at various hospitals, and eventually became a professor at his alma mater. During 11 years of experimentation he accumulated results supporting the concept that nerves and muscle function by electrical mechanisms, which he published in 1791. Galvani discovered that electrical stimulation from many sources can stimulate a motor nerve to cause muscle contraction. Since electrical stimuli give rise to the nerve impulse, Galvani hypothesized that the nerve impulse is itself electrical. The logic of this is not absolutely solid; for example, a spark can ignite a dynamite fuse or the gasoline in an automobile engine. Alessandro Volta (1745–1827), who would go on to develop the first battery, opposed Galvani's conclusion. Volta maintained that the only electricity involved in generating muscle contraction was the electricity introduced from the outside by Galvani's stimulating apparatus.

Somehow Galvani and Volta began to focus their attempts to resolve the issue by examining the electrical phenomena arising from contact between dissimilar metals. It was not known at the time why some combinations of dissimilar metals produced electricity when in contact. Volta hypothesized that even contact between dissimilar nonmetallic substances within the nerve might have produced an unnatural electrical stimulus in Galvani's experiments. In an experiment considered the coup de grâce in defeating this argument, Galvani placed the cut end of the left sciatic nerve of a frog in a position along the right sciatic nerve and likewise placed the cut end of the right sciatic nerve in a position along the left sciatic nerve. This arrangement stimulated both nerves, causing both legs to jump. Galvani's interpretation was that the electric stimulus for contraction was generated internally in the frog with no dissimilar tissues involved that could cause an unnatural effect. But the conceptual framework was wrong. Galvani and Volta and everyone else at the time simply lacked the knowledge of electricity and biology to figure out what was actually happening.

Another misconception in Galvani's time was the apparent lack of insulating material in the body. Without the layer of insulation between two conductors, that is, if two conductors were touching, electric charge spreads out evenly between the two, so they have the same voltage—no difference in voltage between them could be developed or maintained. The bodies of humans and all animals were known at the time to be electrical conductors, so it was hard to see how voltage differences could be maintained in different regions of the body, which would be required for any mechanism involving the flow of electricity. Galvani postulated that the insulating layer must be there but invisible, at the level of the individual nerve and muscle fibers. In 1791 Galvani extracted an oily material from nerve, which he postulated came from the insulating sheet around the central conductive core of the nerve. In this case, postulating the existence of the invisible turned out to be correct, but somewhat for the wrong reasons. Most of the oil he extracted from nerve likely originated from myelin, which is a fatty sheath around fast-conducting nerve fibers in vertebrates. However, many types of nerve fibers and all muscle fibers lack myelin sheaths but conduct impulses just the same. The insulator that plays the fundamental role in animal electricity in all nerves, including myelinated nerves, and in muscle is the cell membrane, and all cells have cell membranes. I wonder if the cell membranes of nerve fibers alone could have supplied enough oily material to be detected in Galvani's analysis. Galvani advanced no theory of nerve conduction beyond his assertion that it is electrical in nature.

Galvani's hypothesis that nerve and muscle are electric machines was correct even though he got many of the details wrong, and it would take a century and a half of further developments in knowledge of the nature of electricity to provide the proof he sought. Galvani published his results in 1791, but sadly his professorship was subsequently taken away because of his refusal to swear loyalty to Napoleon Bonaparte's Cisalpine Republic that ruled Italy. He became a poor man and lived in his brother's house until he died in 1798.

The quest for the explanation of how impulses travel along nerve fibers would not be successfully completed until the mid-twentieth century. Until then there was still significant doubt that nerve impulses, now called action potentials, are fundamentally electrical events. While electrical recordings of action potentials were made in the first half of the twentieth century, the possibility remained that the electrical broadcast picked up by recording electrodes could be an incidental effect of the passage of the signal but not involved in its mechanism. There was reason for skepticism. With few exceptions such as electric organs in a few fish species specialized to produce electric shock, there is no tissue identifiable as an electrical generator in animals. Perhaps the biggest problem was the speed of transmission along nerves, which tops out at velocities near 100 meters per second (224 miles per hour). Although fast enough for distances within a single animal or human body, action potentials travel at nowhere near the light speed of electricity traveling in metal wires that makes telephone conversations over land lines between distant locations possible. Proving that the action potential in nerve and muscle is electrical required the contributions of both the common, but very special, squid that is plentiful along the Atlantic coasts of Britain and America and some sophisticated electronics. Full understanding of how cells produce electricity required a detailed knowledge of how cell membranes and proteins work that was not developed until the second half of the twentieth century.

2

A WORLD OF CELLS, MOLECULES, AND ATOMS

Cells, molecules, and atoms exist in a world of constant motion, which would be a perpetual blur if we could actually look into it with our eyes. The endless motion of atoms and molecules produces the blur. All cells are enclosed within a thin membrane made of fat and protein. The cell membrane harnesses the motion of ions in solution to produce electricity, so atomic theory is crucial to understanding how the body and brain work. The importance of atomic theory to understanding our world was eloquently captured by the physicist Richard Feynman: "If, in some cataclysm, all of scientific knowledge were to be destroyed, and only one sentence passed on to the next generations of creatures, what statement would contain the most information in the fewest words? I believe it is the atomic hypothesis (or atomic fact, or whatever you wish to call it) that all things are made of atoms—little particles that move around in perpetual motion, attracting each other when they are a little distance apart, but repelling upon being squeezed into one another. In that one sentence, you will see, there is an enormous amount of information about the world if just a little imagination and thinking are applied" (Feynman, Leighton, and Sands 1963, 1.2). We start with observations of the unaided human eye, where the jiggling of the particles is hidden, and proceed to the microscopic and submicroscopic realms where it rules.

THE WORLD OF THE UNAIDED EYE

Before the nineteenth century no one had any ideas about atoms, molecules, or cells as we have come to know them. Everything happening inside the human and animal body still had to be explained by plumbing well into the seventeenth century. Inquiries into the function of the nervous system focused on the fluid-filled cavities within the brain—the ventricles (Figure 2.1). There was disagreement about how many ventricles are present, ranging from three to five. Like Galen, we now recognize four, but the more interesting and significant issue was their shape. We now know that the brain and spinal cord develop from the neural tube in the embryo and that the central canal in the spinal cord and the ventricles in the brain are remnants of its lumen. The ventricles and central canal are all interconnected and contain cerebrospinal fluid. In the beginning of the first millennium C.E., there was a great desire to believe that the ventricles are a system of tanks containing animal spirits. However, in dissected brains the ventricles do not look like tanks. They have complex shapes and are more flattened than voluminous, which might suggest that the fluid contained in them is not the main point of the brain, which, of course, it is not. (We now know that the cells lining the ventricles are the important part, some of which are stem cells that can produce new neurons.) The ability of human minds to ignore or twist evidence should never be underestimated, and the theory emerged that the ventricles collapse during dissection as the animal spirits are released. The idea that things may not be what they seem (especially when they go against one's pet theory) was extended to nerves, which were thought to be hollow pipes through which the an-

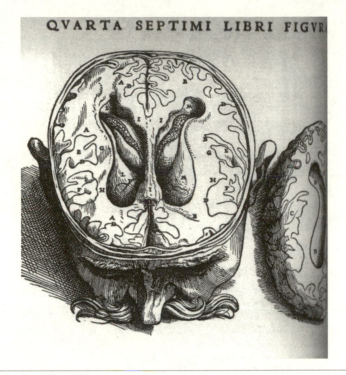

Figure 2.1. Ventricles of the human brain. (From Vesalius 1543.)

imal spirits flowed from the ventricles and into muscles to cause their contraction. The evidence for this was mixed. Blood vessels visible in large nerves could have created the impression that the nerves themselves are hollow, but fine nerves look solid to the naked eye—the only way of looking at them at the time. To get around this, it was hypothesized that fine nerves are also tubes, but with lumens too small to see.

Before the discovery of cells no one could know that nerve impulses are conducted by nerve fibers, now usually called axons, which are fine extensions of neurons. Neurons consist of a cell body from which a number of extensions usually project. The axon is the longest extension and is distinct from all the others, which are called dendrites (Figure 2.2). Dendrites are specialized for receiving signals from other neurons, and action potentials that are generated are sent out along the axon. Just like the cell body from which they extend, axons are filled with cytoplasm, which is fluid containing intracellular structures and covered by the cell membrane. In a sense, axons actually are tubes, but

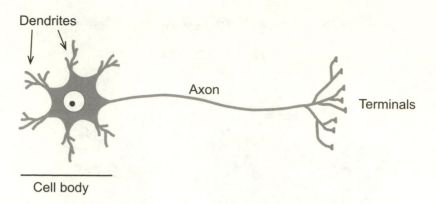

Figure 2.2. External anatomy of a neuron.

the action potential is not a flow of the fluid inside the axon. A nerve is a group of axons assembled into a cable, and each axon conducts action potentials without assistance from the others. Muscle fiber membranes also conduct action potentials, so *action potential* is a much better term to use than *nerve impulse*.

All of that knowledge was yet to come. Previously, an increase in the volume of fluid inside the muscle was believed to provide the power for muscle contraction. The increase in volume was thought to increase the girth of the muscle while pulling the ends closer together, thus generating the force of contraction; that is, the relaxed muscle was long and thin, and the contracted muscle short and fat. The simplest and most direct possible source of the increase in muscle volume was the addition of animal spirits flowing from the ventricles through the nerve and into the muscle. Since it is hard to see how a large-enough volume could be delivered through fine nerves to rapidly inflate the muscle, the idea developed that there is a reaction of animal spirits with substances in the muscle, causing an explosive expansion. The knowledge of plumbing does not seem to have been perfect back then since it seems hard to explain how enough pressure could be produced in the muscle to produce human levels of strength without blowing out the fine nerves connected to them.

Nonetheless, animal spirits remained alive and well into the seventeenth century. René Descartes (1596–1650) was a philosopher and mathematician, not a medical man, and is well known to this day as the inventor of analytic geometry and the Cartesian coordinate system used everywhere, including this book, in the plotting of graphs. His entire life was directed toward academic work. He was a precocious child whose father sent him at age nine to boarding school at Collège Royal Henri-le-Grand, La Flèche, France, where his genius was recognized and personally cultivated by the school's rector. His wide interests included anatomy, some of which he learned firsthand by dissecting animal parts obtained from local butcher shops.

Descartes was highly mechanistic in his thinking, but his ideas about the function of nerve and muscle reached new heights of fantasy. Descartes's work describing the functions of the body, *L'Homme,* written during the years 1629–1633, was published as a section of his masterwork, *Le Monde.* Descartes played it safe by avoiding any comments or conclusions about the soul, and he also delayed publication, to avoid the same fate as Galileo in the hands of the Inquisition. It is also possible that he was a bit of a procrastinator when it came to actually finishing a project. *Le Monde* was published posthumously, more than a decade after his death. He escaped the wrath of the church, but if the church was right about scientists he may not have escaped the wrath of God.

Like Galen, Descartes believed that the ventricles of the brain were tanks from which the tubular nerves transported animal spirits to the muscles. Descartes also believed that animal spirits flowed between antagonistic muscles, the inflow causing contraction, and the outflow relaxation—an interesting twist that mitigated the requirement for a large volume of animal spirits to flow from the thin nerves into the muscle to produce each contraction. Animal spirit theory always had a hard time explaining the function of sensory nerves, since the animal spirits would have to flow toward, not away from, the ventricles. Descartes got around this by promoting the idea that a sensation felt on the body is not communicated by a backflow of animal spirits. Instead, he invented a mechanism in which each nerve contained a filament, essentially a fine thread, extending its entire length, connecting the part of the body feeling the sensation to a valve in the brain. When a sensation is felt, such as the heat of a flame on the finger, the end of the thread inside the nerve is pulled, the valve in the brain opens, and animal spirits flow from the ventricle along the nerve and to the finger just where it is needed, and the finger is withdrawn from the flame—*amazing!*

Descartes did not believe that the heart is a mechanical pump; he thought that the left ventricle contained a substance he called dark fire that ignites and volatilizes the blood, which expands greatly in volume, causing the ventricle to expand and forcing the blood out the aorta. He thought that the warmth provided by dark fire is responsible for the body temperature of warm-blooded animals. It would seem to us easy enough to make the simple observation that blood squirts out of a hole cut in an artery when the ventricle contracts, not when it expands. In fact, one could imagine Galen doing such an experiment more than 1,000 years before Descartes. But Descartes had the mind of a theoretician, not an experimentalist. While Descartes is credited with having the most mechanistic view of the functioning of the brain in his time, to accomplish this he had no hesitation invoking structures and substances that, although not supernatural, were just highly improbable and were never observed, of course, because they do not exist. Descartes's tortuous mechanisms constructed to fit ideas handed down from past authority were not limited to physiology; he developed an extremely contorted theory of earth history in which the earth's crust formed as a shell over a worldwide subterranean ocean. Subsequently the crust broke apart and sank, resulting in Noah's flood (Montgomery 2012). Maybe God was not so upset with Descartes after all.

Francis Bacon (1578–1626), a contemporary of Descartes, is credited with establishing the scientific method as the means to truth. In *Advancement of Learning*, published in 1605, Bacon characterized Aristotle as a "dictator" who, along with his disciples, are "shut up in the cells of monasteries and colleges, and knowing little history, either of nature or time, ... spin out unto us those laborious webs of learning which are extant in their books." He goes on to say that the "wit and mind of man" is usefully applied to study of "the creatures of God," but if it "work[s] upon itself, as the spider worketh his web, then it is endless, and brings forth indeed cobwebs of learning, admirable for the fineness of threads and work, but of no substance or profit" (Smith et al. 2012, 109). One thinks of the fine threads of Descartes that open the valves, letting animal spirits flow into the nerves. In 1660, about three decades after Bacon's death, the newly formed Royal Society codified Bacon's attitude in its motto: *Nullius in verba* (Nothing by words alone). More than any other human organization, before or since, the Royal Society went on to promote and nurture the scientific quest to understand our world (Bryson 2010).

Although Andreas Vesalius (1514–1564) preceded Francis Bacon, he nonetheless introduced full-blown scientific thinking into the field of human anatomy. His teacher during his medical education in Paris, Jacobus Sylvius (1477–1555), later called Vesalius a madman and his major work, *De humani corporis fabrica,* blasphemy. The *Fabrica,* now regarded as the foundation of modern anatomy, had superb illustrations drawn from Vesalius's dissected cadavers (see Figure 2.1), purportedly by the artist Jan Stephen van Calcar, then a student of the great Venetian painter Titian, or in some cases perhaps drawn by Titian himself. Back when he was studying with Sylvius, the only access that Vesalius had to human cadaver material was the bones that he and his medical school buddies clandestinely procured after hangings or lifted from cemeteries. But after Vesalius graduated and ascended to the chair of surgery at Padua, the gallows provided him with a legal supply of hundreds of human bodies.

Galen had coined the term *rete mirabile,* wonderful net, to describe a system of blood vessels at the base of the brain, where he believed vital spirits arriving from the body were transformed into animal spirits before being admitted to the brain. Apparently Galen had only seen the *rete mirabile* in animals because Vesalius could find no *rete mirabile* in human cadavers. It is hard to go against great authorities, so in his medical school demonstrations Vesalius dissected the head of a sheep or ox to demonstrate this structure to his students. As a professor myself, I can easily imagine him explaining that the *rete mirabile* is "harder" to find in humans. He even included a drawing of the "human" *rete mirabile* in a publication that preceded the *Fabrica,* installing into the human brain a structure he had seen only in animals. Such is the power of authority.

Ultimately, Vesalius could not ignore what the cadavers were telling him—the *rete mirabile* just was not there. Moreover, he found the ventricles were filled with a watery fluid in which he could see no magic; the ventricles are portrayed in the *Fabrica* in pretty much their actual shape. His emotional reaction to the strain of being torn between authority and reality is evident in this excerpt from the preface to the *Fabrica.* Leaving no one unoffended, Vesalius wrote, "That detestable procedure by which usually some conduct the dissection of the human body and others present the account of its parts, the latter like jackdaws aloft in their high chairs, with egregious arrogance croaking things they have never investigated but merely committed to memory from the books of others. . . . The former so ignorant of languages that they are unable to explain their dissections . . . so that in such confusion less is presented to the spectators than a butcher in his stall could teach a physician" (Smith et al. 2012, 87). Vesalius mentions no names, but surely Sylvius must have found this statement lacking in the cordiality to which he was accustomed.[1] In this context, Sylvius's calling Vesalius a blasphemer and a madman looks like a proportionate response. It is hard to escape the destiny prescribed by one's personality, and it seems likely that the same personality traits that motivated Vesalius's commitment to truth made it difficult for him to present his results in a more palatable package. It was some time before Vesalius's discoveries were fully accepted into the mainstream of Western medicine.

ENTERING THE UNSEEN WORLD

About a century after Vesalius, when the microscope replaced the naked eye as the main window into the function of the body, ideas began to change. Microscopic observations played a huge role in eventually killing the concept of animal spirits by demonstrating that nerves are not plumbing. Antony van Leeuwenhoek (1632–1723) produced the first images of the nervous system seen through the lens of a microscope. He was the son of a draper in Delft, the Netherlands; obtained no higher education; and followed in his father's footsteps. He was fascinated by the magnifying glasses used in his trade to count threads. The initial fascination turned into action when, during a visit to London in 1668, he examined a copy of *Micrographia* by Robert Hooke (1635–1703), the inventor of the microscope and discoverer of the cell.

Van Leeuwenhoek was a consummate tinkerer, apparently extremely good with his hands, and went on to develop improved microscopes with powers of magnification up to 200 times. He became a regular correspondent to the Royal Society and, in a letter published in 1675, displayed a drawing of a cross section of the bovine optic nerve, which just did not look like a tube. Van Leeuwenhoek discovered that muscles are composed of thousands of fine fibers, and each fiber in turn contained a hundred or so even finer filaments—muscles just did not look like inflatable bags.

The nervous system and muscle revealed by the microscope prompted a reevaluation of ideas about how muscle, nerves, and the brain function. Niels Stensen (1638–1680), the son of a Dutch goldsmith, became a first-class anatomist before eventually joining the priesthood. He had little use for Descartes's theories of nerve and muscle function. Stensen developed a theory in which filaments in muscle that he had observed with the microscope underwent a rearrangement that produced contraction—a harbinger of the modern sliding filament mechanism. Stensen also looked at brain tissue under the microscope. In an admission rare for his time, Stensen wrote that "next to nothing" is known about the brain except that it is composed of gray and white substance (Smith et al. 2012, 125).

It has turned out that the contraction of muscle is a property of individual cells called muscle fibers, each of which extends the full length of a muscle. Muscle fibers contain two types of filaments: one type made of the protein actin and the other type made of the protein myosin. Bundles of actin filaments alternate with bundles of myosin filaments along the entire length of the muscle fiber, with their ends overlapping (Figure 2.3). Successive binding and release of myosin filaments to the actin filaments cause the fibers to slide along one another, greatly increasing the region of overlap, which produces the force of contraction, causing the muscle fiber to shorten (see Chapter 7).

Complementary to Van Leeuwenhoek's refinements of the microscope, a chemical tinkerer, Giorgio Baglivi (1668–1707), developed improved techniques for the preparation of specimens. Baglivi was a physician and a protégé of Marcello Malpighi (1628–1694), who discovered the capillaries that connect arteries and veins. Malpighi also performed some of the first microscopic studies of the brain. The methods he used for specimen preparation included boiling and staining with ink for contrast. These treatments were not conducive to preserving and visualizing structures as they would have appeared in the living brain, but Malpighi did see enough detail to conclude that the cortex of the brain is composed of glands whose ducts head inward in the white matter, delivering their secretions into the ventricles. The secretion was—you guessed it—animal spirits! The fibers in white matter are

a Relaxed muscle

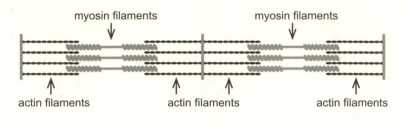

b Contracted muscle

Figure 2.3. Sliding filaments inside a short length of a relaxed and a contracted muscle fiber.

actually axons extending from cell bodies in the cortex. The axons are electrically insulated within many layers of cell membrane called a myelin sheath, which gives the white matter its appearance. However, neurons were not part of the scene in the seventeenth century.

Baglivi achieved some improvement in specimen preparation by soaking his specimens in a variety of solutions including wine, alcohol, acids, and milk. Trial and error became the name of the game in preparing specimens for microscopy and remained so well into the twentieth century. The chemical basis of specimen preparation is now fairly well known. Preservatives such as formaldehyde cause chemical bonds to form within protein molecules preserving their shape and between protein molecules in proximity to one another, binding the cells and extracellular fibers in the positions that they held in life. Then, dyes that have specific staining characteristics can be applied; for example, hematoxylin and eosin mixed together stains positively charged proteins in the nucleus of the cell blue and stains other structures in cells and fibers various shades of red, orange, and pink, together producing an appealing and therefore convincing picture.

While Baglivi did not reach the level of visualization of microscopic specimens achievable today, he was able to see the fibrous structure of muscles, tendons, and bones. (Bones are composed of fibers made up of the protein collagen upon which calcium has been deposited.) He also visualized fibers in blood vessels, glands, nerves, most internal organs, and the membranes covering the brain and spinal cord. Among Baglivi's interesting "observations" was that fibers are softer in women and children than in adult men and that the fibers in the French are softer than the fibers in Italians, which are in turn softer than the fibers in Spaniards and Africans (Smith et al. 2012, 179). Who would have guessed?

Influenced by his observations of the intricate fiber structure of muscles, Baglivi began to believe that the muscles themselves generated the force of contraction, rather than being powered by the pressure of animal spirits delivered by the nerve: "having often and carefully examined the structure of the muscles . . . I began to assert that the principal, not to say the whole power of motion, or the force moving the muscles . . . resides in the muscles themselves, that is, in the special fabric of the fibers . . . and that the spirits flowing through the nerves serve no purpose other than to regulate their motions" (Smith et al. 2012, 180). In drawing this conclusion, Baglivi was dead on, but he had his own fanciful theories about how the fibrous covering of the brain contracted and squeezed the animal spirits into the nerve and to the muscle to trigger their contraction. So neither plumbing nor wild speculation were dead, but the notion that tissues such as muscles were autonomously capable of carrying out their functions, rather than just effectors powered by the enigmatic brain, was a transformation in thinking—not to mention that it also happens to be true.

Perhaps the greatest paradigm shift in science ever came when John Dalton (1766–1844) proposed his atomic theory of 1803 that the world is made of atoms that combine together in fixed proportions to form molecules; for example, two atoms of hydrogen (H) combine with one atom of oxygen (O) to form a water molecule (H_2O). This idea laid the groundwork for the modern scientific view of the world. Dalton's atomic theory did not take hold at once because it could not explain exceptions such as mixtures of gases, mixtures of substances in solution, and mixtures of metals to form alloys, all of which occur in variable proportions (Coffey 2008). Ultimately, the distinction between atoms that combine together chemically, in proportions dictated by the number of bonds an atom can make, and simple mixing in variable proportions without chemical bonding was recognized, and atomic theory became part of the foundations of modern chemistry and physics.

Without atomic theory, Svante Arrhenius (1859–1927) could never have discovered ions. When he arrived on the scene, atomic theory was well established, and Demitri Mendeleev (1834–1907) had arranged the known atoms in the periodic table, but atomic structure was unknown and atomic masses had not been determined. Arrhenius was a disobedient graduate student viewed as lazy and not highly skilled in the laboratory by his two supervisors at Uppsala University in Sweden (Coffey 2008). He pursued an independent project for his thesis, against their wishes, which is sometimes the best way to go, especially for a creative mind. But in the short term this was not an optimal career decision, and his poor grade on his doctoral thesis initially disqualified him from obtaining a university teaching position. After obtaining his third-rate doctorate (not just a casual slur—doctorates were actually assigned grades), Arrhenius pursued his research while unemployed, living for several years off an inheritance from his recently deceased father, who had been on the support staff of Uppsala University. Then as now, being tall and physically attractive had its advantages, but apparently Arrhenius was neither. Wilhelm Ostwald, who would become an important mentor to Arrhenius, described Arrhenius after their first meeting in a letter to his wife as "somewhat corpulent with a red face and a short moustache, short hair: he reminds me more of a student of agriculture than a theoretical chemist with unusual ideas" (Coffey 2008, 17). Nonetheless, the two became coworkers and friends, and Arrhenius ultimately had a successful career. He secured a faculty position at Stockholm University College, helped to set up the Nobel Institute and the Nobel Prizes in 1900, and received the Nobel Prize in Chemistry himself in 1903.

Of his many contributions, the one most relevant to our subject is the ionic theory of solutions, which Arrhenius began to develop in his doctoral dissertation. As happens in so many research projects, Arrhenius's initial goal was quickly set aside as the results of the first experiments led him in a different direction. Initially, Arrhenius was attempting to determine the relative number of sugar molecules added to a salt solution by determining the effect of added sugar on the electrical conductivity. Since this project went nowhere, we can skip the rationale. Arrhenius's method turned out not to work very well, and the much better approach of determining how much the addition of the substance in question depresses the freezing point of water (which is how antifreeze works) had just been developed by François Raoult in 1882, but Arrhenius was unaware of this. Freezing-point depression became the standard method of determining the relative number of atoms or molecules added to the water and thus the atomic or molecular mass.

In the latter years of the nineteenth century when Arrhenius made his observations, electrons and protons were yet to be discovered, so the nature of the charged entity or entities that carries electric current between two electrodes immersed in a solution of an electrolyte in water was unknown. An electrolyte is any substance that conducts electricity when dissolved in water. Electrolytes include salts (e.g., sodium chloride), acids (e.g., acetic acid), and bases (e.g., sodium hydroxide). At the time it was difficult to believe that any electrolytes actually dissociated very much into component parts when dissolved in water. For example, the formation of sodium chloride salt from pure sodium metal and chlorine gas releases a large amount of energy in the form of heat. Thus, it seemed that an equivalent large amount of energy should have to be supplied to separate sodium and chloride in salt from each other, and it seemed impossible that this could happen just by dissolving salt in water. Where would the energy needed to break them apart come from? No one had any idea that a much smaller amount of energy was needed and could be supplied by the water molecules themselves, as is now known to be the case (see below). In the nineteenth century it was more reasonable to think that sodium and chloride stayed together in solution and somehow picked up the charge entering the solution from the positive electrode and passed it along bucket-brigade style, delivering it to the negative electrode. Arrhenius found that in the case of sodium chloride and other strong electrolytes, the conductivity of the solution was proportional to the amount of electrolyte added; double the electrolyte added and the conductivity doubles, presumably because the number of particles available to carry electric current is doubled. This is unremarkable and provides no information about whether sodium and chloride remain together or completely dissociate in water. In either case, doubling the sodium chloride added would be expected to double the number of particles available to carry electric current.

The breakthrough came when Arrhenius discovered an anomaly in the case of weak electrolytes. When acetic acid, a weak electrolyte, was mixed into water at low concentrations, the proportionality of concentration and conductivity was just the same as in the case of a strong electrolyte such as sodium chloride. However, as the concentration of acetic acid added to water was increased, the corresponding increases in conductivity became progressively much smaller than was produced by adding the same concentration of sodium chloride. It was as if adding more acetic acid to water already containing a high concentration does not increase the number of particles that carry current nearly as much as when the same amount of acetic acid is added to pure water. The way Arrhenius figured that this could happen is if the acetic acid molecule itself is incapable of carrying electric current and that in water it dissociates into component particles, which are the real current carriers.

We now know that the energy required for dissociation is supplied by the association of the component particles with the water molecules. Weak electrolytes are weak because a larger amount of energy is required to dissociate a weak electrolyte into its component particles than is required to dissociate a strong electrolyte. Because of this, it is harder for water to supply enough energy to dissociate the weak electrolyte into its components, and the component particles in solution tend to bind together again, releasing their energy when they bump into each other. Strong electrolytes requiring less energy just dissociate into their component particles in water, and that's it.

Arrhenius figured out that at high concentrations a weak electrolyte such as acetic acid might be expected to dissociate only incompletely because the dissociated particles would frequently collide with one another and bind together, causing reassociation to occur, re-forming an acetic acid molecule. This does not tend to happen at low concentration because the dissociated components of a weak electrolyte are so dilute they practically never find each other. Thus, at high concentration of a weak electrolyte equilibrium might be established in which both dissociation and reassociation are occurring simultaneously, and only a fraction of the acetic acid molecules are dissociated into current-carrying particles at any given time. The exact fraction would be determined by the concentration of acetic acid that was added to the water. (We now know that an acetic acid molecule dissociates into a negatively charged acetate ion and a positively charged hydrogen ion.)

Arrhenius was able to make a few friends after obtaining his doctorate, and his association with Jacobus Henricus van't Hoff (1852–1911), who was investigating osmotic pressure as a way of making relative counts of atoms and molecules in solution, provided the key to determining that electrolytes in water really do dissociate into component parts. Van't Hoff's insight was that osmotic pressure in liquids could be explained by the same kinetic theory that had been developed for gases. In the kinetic theory of gases it was postulated that the molecules occupy a tiny part of the full volume of a gas. They zip around in perpetual motion propelled by heat energy, and when contained within a closed vessel, the collisions of the gas molecules with the walls of the container produce pressure. An increase in temperature increases the velocity of the molecules, which in turn increases both the frequency of collisions and the energy of collisions with the walls, and thus increases the pressure. If the temperature remains constant, squeezing a gas into a smaller volume, such as with a piston pushed into a closed-ended cylinder containing the gas, increases the frequency of collisions with the piston and the walls of the cylinder and thus increases the pressure. Boyle's law, formulated in 1662, states that at a constant temperature the volume and pressure of a given amount of gas are inversely proportional—double the pressure and the volume of the gas collapses to one-half, reduce the pressure to one-half and the volume of the gas doubles.

The gas-like behavior of particles dissolved in water is evident when osmotic pressure develops across a semipermeable membrane. Many kinds of semipermeable membrane exist, the simplest of which is a sieve-like membrane penetrated by pores that water molecules can easily pass through, but larger molecules such as sugars cannot. Although molecules in a liquid are closer together than molecules in a gas, they still zip around banging into each other and the walls of their container. To understand how osmotic pressure works, let's perform a thought experiment in which two containers are connected together by a pipe at their bases with a partition made of this kind of semipermeable membrane in the middle (Figure 2.4). Water can readily pass through the pores in the membrane, but sugar cannot.

The containers start out with equal volumes of water, and the water molecules are continuously colliding with the container walls. When water molecules collide with the semipermeable membrane at a pore, they just pass through as if the membrane was not there. When both containers hold pure water, the water flow across the membrane left to right equals the water flow right to left, so the net flow is zero, and the

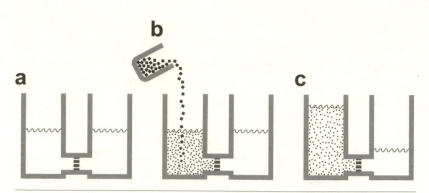

Figure 2.4. Osmotic pressure developed when sugar granules are dissolved in water.

water levels in the containers remain equal (Figure 2.4a). But the addition of sugar to the left container changes the scenario (b, c). Sucrose molecules collide with the membrane but cannot pass through because the pores are too small, and the presence of the sucrose molecules in the left container reduces the frequency that the water molecules in the left container strike the pores, thus reducing the flow of water molecules from left to right. However, the presence of sucrose in the left container does not reduce the frequency of collisions of water molecules in the right compartment with the pores in the membrane, and thus does not much interfere with the right-to-left flow of water through the pores. Thus, the presence of sucrose in the left container results in a net flow of water from the right container into the left container. This net flow of water causes the water level in the right container to fall and the level of the sucrose solution in the left container to rise.

The weight of the portion of the sucrose solution in the left container that is above the water level in the right container produces pressure that drives water through the semipermeable membrane back into the right container. The fluid in the left container rises until the pressure difference reaches a level that exactly counteracts the flow of water into the left container due to the sucrose, and then the fluid levels stabilize. The pressure at which that occurs is a measure of the osmotic pressure exerted by the sucrose. How much osmotic pressure is developed depends upon the concentration of sucrose molecules in the left container—double the sucrose concentration and the osmotic pressure doubles. So even if you have no idea how many sucrose molecules are in a solution, if you add just the right amount of sucrose to double the osmotic pressure, then you know that the number of sucrose molecules has doubled.

In Arrhenius's time semipermeable membranes were available with pores that allowed the passage of water molecules but not sodium chloride, so the relative number of "molecules" of sodium chloride in a solution could be measured by the osmotic pressure method. Arrhenius found what he had suspected: a solution of sodium chloride produced twice the osmotic pressure that would be produced if sodium chloride dissolved in water remained as intact molecules. The obvious conclusion: when dissolved in water, sodium and chloride completely dissociate into separate "atoms," which are the current carriers. This was a momentous discovery that led to the conclusion that salts like sodium chloride are electrically conductive because they dissociate in water into electrically charged forms of the atoms, which became known as ions. In the case of sodium chloride the ions are, of course, positively charged sodium ions and negatively charged chloride ions. Thus, it is the ions that carry the current in solutions that conduct electricity. Pure water, in fact, produces its own ions by the dissociation of a very small fraction of the water molecules into hydrogen ions (H^+) and hydroxyl ions (OH^-), which confers a small electrical conductivity on pure water, albeit much lower than the conductivity of seawater or body fluids, which contain dissolved ions.

There are a few points about ions and semipermeable membranes that merit elaboration. Semipermeable membranes that exclude the passage of sodium and chloride as used by Arrhenius exclude these ions from their pores because of the electric charge of the ions. Sodium and chloride ions themselves are of a size similar to that of individual water molecules, so there is no large size difference as is the case between water and sugars such as sucrose. But the ions in solution are surrounded by a shell of water molecules that are bound to them electrically (see below), which effectively makes them considerably larger than individual water molecules and so excludes them from entering the pores of the membrane.

Now, why isn't a prohibitively large amount of energy required to dissociate sodium chloride into individual sodium and chlorine particles when sodium chloride dissolves in solution? While a tremendous amount of energy is liberated when sodium metal and chlorine gas are combined to make sodium chloride, it is the transfer of electrons that produces the ions that releases the energy. The sodium and chloride ions in salt stick together because of the electrical attraction of their opposite charges. Therefore, separation of sodium and chloride ions in solution, which happens when salt is dissolved, is not the reverse process of restoring the sodium and chloride to the uncharged form they had in sodium metal and chlorine gas and does not require the input of such a large amount of energy. It turns out that the energy that is required to dissolve salt into ions is supplied by the interaction of the charges on the ions with partial electric charges on the water molecules (see below). So the original bias against the possibility that sodium and chloride dissociate when dissolved in solution because too much energy would be required is easily dismissed in light of modern chemistry.

Semipermeable membranes and osmotic pressure are not just the playthings of scientists. Actually semipermeable membranes are the basis of all cellular life. Cell membranes evolved because they confine the proteins, ions, sugars, and other molecules required for life processes inside the cell and exclude interfering molecules. Mechanisms evolved that prevent the cell from swelling and exploding from the water drawn in by osmotic pressure. The resulting unequal distribution of ions on each side of the cell membrane produces an electrical charge across the cell membranes of all cells, including plant cells. In some cells, such as neurons and muscle cells, the electric charge across the membrane provided the basis for further evolution into animal electricity and all of its consequences, such as sensation, movement, brain function, and consciousness. Arrhenius, the problem student who was awarded a third-rate doctorate, really discovered something fundamental when he developed the ionic theory of solutions.

It was in 1897, shortly after Arrhenius deduced the ionic theory of solutions, that a team of British physicists led by J. J. Thomson discovered the electron. This was accomplished by investigations of cathode rays. The cathode ray is a wonderful and transformative invention—nothing less than tamed lightning! First the necessary definitions: *cathode* is another name for a negative electrode, and *anode* is another name for a positive electrode. Production of a cathode ray is accomplished by configuring the anode as a disk with a small hole in the center placed like a target in front of the cathode. When a cathode and anode are enclosed in a vacuum tube without touching each other, and the cathode is heated to a high temperature, electrons are emitted from the cathode and fly in a straight line toward the anode, being attracted by its positive charge. (There has to be a vacuum inside the tube or else the whole system would burn up.) Each electron is attracted to the anode equally in all directions as it approaches the hole in the anode, so it has no choice but to pass right through the middle of the hole. Carried by momentum, the electrons continue forward until they collide with the surface of the tube, which is coated with phosphors that glow when electrons strike them. The tube surface is also an anode, and after striking it the electrons are conducted away.

This arrangement was utilized in the development of the cathode-ray oscilloscope, which revolutionized how voltage is measured. A pair of lateral plates, one on each side of the electron beam and located between the anode and the screen, is supplied with a smoothly changing voltage, which deflects the beam across the screen from left to right at a constant rate. The result is a glowing point that travels from left to right across the screen, or, if the left-to-right deflection is fast and repeated at a high rate, a bright horizontal line is produced across the screen. The rate can be made very fast, creating a millisecond time scale. When a voltage is applied to a pair of plates mounted vertically above and below the horizontal beam, the beam moves up or down according to the applied voltage as it traverses the screen from left to right, tracing the change in voltage with time. The cathode-ray oscilloscope was the state of the art for measuring the electrical activity of nerve fibers for decades (see Chapter 3).

The cathode-ray tube provides an excellent example of unanticipated spinoff of a scientific development that profoundly changed our world. For the half century before the development of flat-screen displays, the cathode-ray tube was the only way to go to produce a television picture. In the case of the black and white TV picture tube, vertical plates were supplied with a smoothly changing voltage that moved the beam from top to bottom at a constant rate. Simultaneously, an alternating voltage was applied at a much higher frequency to the lateral plates that rapidly flicked the beam from left to right, over and over again, while the intensity of the beam was varied to produce lighter and darker regions according to the signal fed into the TV amplifier. The result was that the cathode ray "painted" a picture on the screen extremely fast. The entire process took place repeatedly at a higher frequency than can be resolved by the eye, producing apparently seamless moving images on the TV screen.

In 1909–1913, long before the development of these applications and just a few years after the discovery of the electron, Robert Millikan and his student Harvey Fletcher used cathode rays to accomplish the formidable task of measuring the charge of a single electron (Johnson 2008). Millikan projected a beam of electrons into an aerosol of oil droplets suspended in a vacuum chamber. Some of the oil droplets became negatively charged, indicating that electrons had stuck to them. The oil droplets slowly fell due to the force of gravity, and Millikan found that a vertical electric field created between a negatively charged plate at the bottom of the chamber and a positively charged plate at the top of the chamber could be adjusted to a voltage that exactly counteracted the force of gravity on some of the droplets, preventing their fall. This was a painstaking experiment. Millikan and Fletcher looked through a horizontally mounted microscope, targeted a single oil droplet, and then adjusted the voltage on the plates to exactly counteract its fall, and wrote the voltage down in their lab notebook—over and over again. To other scientists this began to look like an obsessive-compulsive disorder.

The results were stunning. The force imparted by the electric field on an oil droplet depends upon the number of electrons sticking to it. Millikan found that the electric field required to float oil droplets of the same size did not vary smoothly but in discrete steps, with some droplets requiring an electric field of maximum strength, some requiring half the maximum field strength, and some one-third of the maximum field strength, and so forth. Since the force of an electric field is felt equally by each electron, an oil droplet would be buoyed up in proportion to the number of electrons stuck to it. Since electrons add virtually no weight, to be supported against gravity an oil droplet bearing two electrons would require one-half of the electric-field strength of an oil droplet bearing one electron, an oil droplet bearing three electrons would require one-third of the field strength of an oil droplet bearing one electron, and so forth. From careful measurements of the electric-field strength required to float charged oil droplets, Millikan was able to determine the electric charge of a single electron to within 1% of the currently accepted value of 0.00000000000000000016021765 coulombs, a very small number indeed! The tiny number produced by the obsessive-compulsive behavior of Millikan secured him the Nobel Prize in Physics in 1923.

The number of coulombs of charge on a single electron can just be flipped (and rounded a little) to reveal that the charge of 6,240,000,000,000,000,000 electrons adds up to one coulomb of negative charge. Those of us who are scientists or engineers would never write out a large number like this because it is hard to deal with and meaningless to verbalize—does 6.24 quintillion have any meaning for you? Expressed in scientific notation the same number becomes 6.24×10^{18} and is much easier to deal with. If your memory of scientific notation is nonexistent or rusty, see the endnote.[2]

The Millikan oil-drop experiment made single electrons "visible" by attaching them one at a time to objects that could be seen with a microscope. Once "visible" the movement of single electrons in an electric field could be studied. Botanist Robert Brown (1773–1858), while looking through a microscope at pollen grains floating in water, noticed that they jiggled around in constant random motion. This simple observation was so important that it was named "Brownian motion" in his honor. I remember seeing Brownian motion for real through a microscope during a high school biology class. Brownian motion is a phenomenon where the atomic nature of things breaks through, as if begging to be recognized. Pollen grains are light and are bounced around by the constant motion of water molecules. It takes a microscope to see it because objects big enough to see with the naked eye are too massive relative to water molecules to be noticeably bounced around by molecular collisions, but once the microscope was developed it was obvious.

Jean Baptiste Perrin (1870–1942) devised a way to bring atomic-scale particles into the visible world such that their actual weight could be determined, and thus the number of atoms or molecules in any specified weight of a substance could be counted (Perrin 1926)—an amazing capability. (A word is in order about the terms *mass* and *weight*. They are frequently used interchangeably, but weight depends on the magnitude of the force of gravity and is only equal to mass when measured on the surface of the earth.) Perrin's method involved creating an experimental system where the effect of Brownian motion could be measured. Since the magnitude of Brownian motion of an object or particle depends on its mass, measuring Brownian motion provided the key to finding the actual mass of all atoms and molecules.

The method involved a suspension in water of microscopic spheres of uniform size made of resin. The resin used in many of the experiments was gamboge, a mustard-yellow dye traditionally used to color the robes of Buddhist monks. The dye consists of spherical particles of assorted sizes from which particles of the desired uniform size could be obtained by repeatedly centrifuging a suspension of the particles to remove the heavier and lighter ones. Centrifugation is a standard technique in biochemistry and biology. Tubes filled with fluid containing the particles of interest are loaded onto the rotor of a high-speed centrifuge and spun at high revolutions per minute, resulting in the heavier particles settling to the bottom and floating the lighter particles on top. The fraction of the solution in the tubes containing the particles that include those of the desired weight is collected and then centrifuged again to achieve further purification until only particles of the desired size remain. The process is repeated as many times as it takes to achieve the desired purification. This is painstaking work. A kilogram of dye yielded just a few tenths of a gram of gamboge particles, each about 0.75 micrometers in diameter. A micrometer (μm) is one thousandth of a millimeter (mm). The gamboge particles used were a little smaller than the diameter of the finest nerve fibers, but nowhere near molecular dimensions, and easily visible with a compound microscope.

Gamboge particles mixed in a container of water sink to the bottom, but being small enough to be kicked around by Brownian motion, they do not just settle down and lie there. A fraction of the particles are always suspended above the bottom, forming a visible haze. Scientific creativity often involves seeing the same principle underlying apparently distinct phenomena. Perrin recognized that the haze of gamboge particles was produced by the same principle that is responsible for the molecules in air being suspended in the atmosphere rather than just settling to earth and lying on the ground. The gravitational field of the earth attracts the molecules in the atmosphere, just as it attracts gamboge particles. They would fall to the ground were it not for their constant random motion that keeps them airborne.

The density of molecules in the atmosphere and the density of gamboge particles suspended in water decrease as the altitude above the surface increases. It turns out that both decreases in density are mathematically related to particle mass. Perrin realized that if he could determine the relationship between particle mass and the decrease in density with altitude for gamboge particles in suspension in water, then the actual mass of any molecule in the atmosphere could be calculated from its decrease in density with altitude.

The mass of individual gamboge particles in a sample containing particles of uniform size can be simply determined by weighing a sample containing a known number of particles on a sensitive scale. An increase in altitude of 5 kilometers above the surface of the earth results in a 50% reduction in the density of oxygen molecules. To determine the increase in altitude within the haze of gamboge particles, Perrin sealed a suspension of gamboge particles in water on a microscope slide with a glass coverslip, producing a transparent chamber about 0.1 mm deep. He used a microscope equipped with a lens with narrow depth of field limited to an extremely thin plane above the surface of the slide. He recorded the number of particles in the field of view that were in focus, then he refocused at a higher plane and recorded the number of particles in focus and the distance above the initial plane of focus. In this way he was able to determine that a 6-μm change in altitude produced a 50% reduction in the density of gamboge particles. With knowledge of the mass of the gamboge particles, and the changes in altitude that produced 50% reductions in gamboge sphere density and oxygen density, Perrin determined that a single oxygen molecule has a mass of about 5.36×10^{-23} grams.

This is the actual mass of the oxygen molecule, not to be confused with the concepts of atomic mass or molecular mass, which are relative quantities. Until Perrin's measurement, the mass of any atom or molecule was known only in these relative terms. Since the mass of an atom or molecule overwhelmingly resides in the protons and neutrons in the nucleus, and since protons and neutrons have very nearly equal masses, the atomic mass is very nearly equal to the number of protons and neutrons. Thus, an oxygen atom has an atomic mass of 16 because it contains eight protons and eight neutrons. A hydrogen atom, being a single proton, has an atomic mass of 1. Oxygen molecules contain two oxygen atoms, so the molecular mass of an oxygen molecule is 32. Likewise, hydrogen molecules contain two hydrogen atoms and have a molecular mass of 2. Since atomic and molecular masses are relative quantities, an oxygen molecule weighs 16 times as much as a hydrogen molecule.

There is a practical consequence of this, which has been extremely important to the development of the field of chemistry. If we have an amount of oxygen weighing 16 times the weight of an amount of hydrogen, both contain the same number of molecules. The absolute amounts and the units of weight that are used make no difference; as long as the amount of oxygen weighs 16 times the weight of the amount of hydrogen, both contain the same number of molecules. The unit of weight and mass that has been chosen for use in science is the gram, and the atomic or molecular mass of a substance with the gram unit attached is called the gram-atomic mass (e.g., 16 grams for the oxygen atom) or the gram-molecular mass (e.g., 32 grams for the oxygen molecule). The gram-atomic mass or gram-molecular mass of all pure substances contains the same number of atoms or molecules; this number is called a mole. Thus, the number of moles of a substance can be determined just by weighing it.

Once Perrin had determined the actual mass of an oxygen molecule, he calculated that 32 grams of oxygen, that is, one mole of oxygen, contains between 6.0×10^{23} to 6.8×10^{23} molecules. The number of atoms or molecules contained in one mole is the same for all substances and was named "Avogadro's number" in honor of Amedeo Avogadro (1776–1856), whose gas law, that equal volumes of all gases contain the same number of molecules, had pointed to the atomic nature of all matter in 1811. Modern methods pinpoint the mole as containing 6.02×10^{23} atoms, molecules, or ions. So the gamboge method, while a bit less accurate, was essentially correct! Of course, Perrin had determined not just the actual mass of oxygen, but of all atoms and molecules whose atomic or molecular masses were known or would be known in the future. It was of immense importance. Whenever I recall Perrin's experiment I am amazed by the imagination it must have taken to think that such an experiment could work.

Moles per liter is the unit of concentration of particles in solution, and a solution containing 1 mole of a substance is referred to as a 1 molar solution. The number of ions in a mole turns out to be almost 100,000 times greater than the number of charged particles in a coulomb; there are 96,485 coulombs of charge per mole of electrons or protons. This value is called the Faraday constant. Thus, a mole of sodium ions or potassium ions each contains 96,485 coulombs of charge. The mole and Faraday's constant are important concepts for animal electricity because the animal battery generates a voltage by virtue of different concentrations of sodium and potassium in the fluids inside and outside of the cell (see Chapter 3). Concentration gradients depend on differences in the number of particles inside and outside the cell, so moles per liter is the relevant unit. Faraday's constant allows the concentration differences to be expressed in electrical terms in the Nernst equation. The voltage developed by the animal battery drives all of the ionic currents that are the basis of the functions of our sensory systems, the nervous system, muscle, and a whole lot more.

THE CELL MEMBRANE

This chapter began with consideration of some of the conclusions drawn by early investigations of the structure of human and animal bodies. We have worked our way down through the microscopic scale to the scale of atoms, molecules, and ions. The view at the microscopic scale reveals much about what is really there (cells and their products) and what is really not there (tubes that carry animal spirits). The view at the atomic and molecular scales reveals the world of Brownian motion where everything that happens is influenced by the constant collisions of atoms, molecules, and ions with one another, driven by the thermal energy all around us. Now let's start within the world of Brownian motion to construct a picture of the cell as a generator of electricity.

This construction project begins with arguably the most wondrous molecule of all—water. The oxygen nucleus in a water molecule consists of eight protons and eight neutrons, and it is much larger than the hydrogen nuclei, which have one proton each (Figure 2.5). Water is

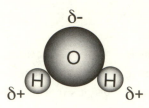

Figure 2.5. Depiction of a water molecule with partial positive (δ+) and partial negative charges (δ–) indicated.

electrically neutral since the electric charge of the water molecule's 10 electrons exactly neutralizes the electric charge of its 10 protons. However there is a slightly greater attraction (called electronegativity) of the larger oxygen nucleus for some of the electrons, which results in the oxygen nucleus hogging a little bit more than its fair share of their attention. This produces a partial negative charge associated with the oxygen nucleus, leaving behind an equal partial positive charge shared evenly between the two hydrogen nuclei. Water and other molecules that have partial positive and partial negative charges in different regions of the molecule are called polar molecules, because of their charged "poles." In liquid water, the partial negative charge of the oxygen pole of each water molecule is repelled by the oxygen poles and attracted by the hydrogen poles of its neighbors. This imparts a structure to liquid water where the water molecules orient such that the oxygen nuclei of the water molecules are nearer to the hydrogen nuclei than to the oxygen nuclei of their neighbors. The polar feature of water molecules is what makes water the essential medium for life. It is why looking for life on other planets involves looking for evidence of water. Water is not just something to drink. You will soon see why life just doesn't happen without water.

Salt dissolves in water because of the polar nature of the water molecules. When a crystal of solid salt such as sodium chloride is immersed in water, the oxygen poles of the water molecules are attracted to the positive charges of the sodium ions and the hydrogen poles are attracted to the negative charges of the chloride ions on the surface of the salt crystal. Constantly battering the salt crystal with their thermal motion, the water molecules eventually insert themselves between the sodium and chloride ions. In effect, the water molecules pry the sodium ions and chloride ions off the surface of the salt crystal using the energy supplied by the interaction of their partial charges with the charges of sodium and chloride in the crystal. Because the charges on individual water molecules are partial, several water molecules must interact with the salt crystal to liberate each sodium ion and each chloride ion into solution. In scientific terms, the energy required to break an ion loose from the surface of the salt crystal is supplied by the energy released when the ion binds with the water molecules.

Once free in solution, the ions remain tightly associated with the water molecules that liberated them. Six water molecules surround each dissolved sodium ion arranged in a cage called a hydration shell, with walls one water molecule thick. The negative oxygen poles of the water molecules in the shell face inward toward the captive sodium ion. The water molecules in the hydration shell also interact with one another, and their hydrogen poles cannot just snuggle up together on the outer surface of the hydration shell. This forces the water molecules to lock into a structure where their oxygen poles simultaneously interact with the sodium ion and the hydrogen poles of the neighboring water molecules. The hydration shell formed by these interactions is a rigid structure with a cavity inside that is larger than it needs to be to contain the sodium ion, but fewer water molecules cannot form a shell with enough space, so the sodium ion gets room to spare. Potassium ions, although about 20% larger in diameter than sodium ions, are also accommodated within an identical shell of six water molecules. Chloride and other negative ions also have hydration shells, in which case the water molecules in the hydration shells arrange themselves with their positively charged hydrogen poles facing the negatively charged ion.

The electric field extending around a sodium ion extends beyond the water molecules in the hydration shell. Although polar, water molecules are electrically neutral, so a sodium ion together with all the water molecules within range of its electric field still has the same single net positive charge as a sodium ion alone. Although water does not neutralize the charges of the ions dissolved within it, by aggregating around each ion the water molecules force the ions to keep their distance. In effect, the water molecules provide lubrication that allows ions of opposite charge to slide past one another. How much water the ions take with them as they carry electric current is still a matter of research. There are a number of additional loosely arranged hydration shells layered on top of the first shell encasing the ion. Imagine a sodium ion streaking through the water, within the cage formed by its first hydration shell, while dragging a loosely attached plume of water molecules along, so that some of the water is continually lost from the tip of its tail and is replaced with other water molecules encountered along its path.

An electric field imposed upon a sodium chloride solution causes sodium with associated water molecules to flow in one direction and simultaneously causes chloride with associated water molecules to flow in the opposite direction. Unlike electrons moving through a wire, which are always neutralized by the charges of the nearby stationary protons fixed within the structure of the metal, sodium and chloride ions achieve a similar neutralization of each other's charge during their passage by one another as they move in opposite directions carrying the current. All the charged particles in an ionic solution are movable. The solutions inside and outside the cells contain sodium, chloride, potassium, calcium, and other ions, all of which participate in carrying current by moving through the cells and extracellular fluid.

At some remote time and place in the deepest reaches of the history of life on earth, the cell membrane evolved. No one knows whether life formed first as an aggregation of self-reproducing molecules and then acquired a membrane around itself or whether a primitive cell membrane formed spontaneously and then the processes of life developed within it. But there is no disagreement about the tremendous importance of the cell membrane and the other membranes within the cell to the complex life forms that inhabit the earth. The cell is the unit of function of all complex life processes. While some cellular functions, such as the contraction of muscle fibers in skeletal muscle, are accomplished by many cells that have fused together sharing a single cell membrane, these cells have simply joined forces, enabling them to perform their function as a larger cell of a shape and size better suited to their task.

The common aphorism "oil and water do not mix" captures the fundamental principle underlying the structure of the cell membrane. Phospholipids that make up the bulk of the cell membrane are not chemically bound together. They simply would rather be together with each other than on their own among the water molecules. This is why it is plausible that primitive cell membranes enclosing tiny volumes of water may have formed spontaneously within the primordial ooze. The cell membrane holds together because of the same simple physical principle that causes salt to dissolve in water—unlike charges attract each other and like charges repel each other. Fats and oils are largely composed of molecules without regions of full or partial charge; unlike water molecules, they are nonpolar. Oil and water do not mix because nonpolar molecules cannot intermingle with polar molecules that are attracted to each other's partial negative and positive charges—the immense electrical force prevents it.

Phospholipids, the fat molecules that form the basic structure of the membrane, partially defeat their exclusion from water because each phospholipid molecule consists of a head region that is electrically charged and so interacts with water and two nonpolar lipid coattails that are neither charged nor polar and so are excluded from water. Figure 2.6 shows the progression from the phospholipid molecule (a) to the cell membrane (e). Lapsing into anthropomorphism, scientists refer to the charged head regions that interact with water as hydrophilic (water loving) and the nonpolar tails that do not interact with water are hydrophobic (water hating). Without the hydrophilic heads, the lipid molecules would coalesce into floating blobs like drops of cooking oil in a pot of pasta water. Instead, individual phospholipid molecules (a) line up with their hydrocarbon tails side by side, minimizing their contact with water, and with their charged or polar heads facing outward forming an interface with the water. Since the tips of the tails do not like to touch water either, one of the most stable configurations of phospholipids in water, called a lipid bilayer, consists of a sheet of two layers of phospholipid molecules with their tails facing inward toward each other and their heads facing outward interacting with water on both sides of the sheet (b). Cell membranes are ordinarily represented in diagrams by two parallel strips that represent the hydrophilic head groups with a lighter region in between representing the hydrophobic tails (c). When such a membrane is continuous and entirely encloses a volume within a watery solution, no lipid tails are exposed to water at their lateral edges or tips, and the membrane forms a stable structure maintained by the interactions of the charged head groups with water inside and outside the space enclosed by the

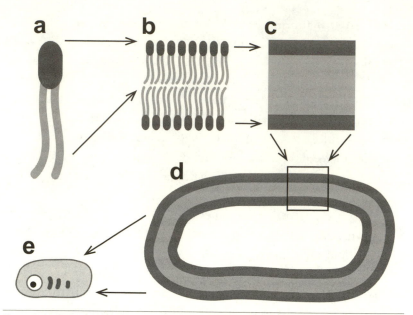

Figure 2.6. The phospholipid structure of the cell membrane. Note the variations in scale.

membrane (d). Add a nucleus and some other stuff and—voilà—we have a cell (e).

Many different proteins specialized for different functions reside among the phospholipids in cell membranes. Proteins are the Legos of biology, consisting of amino acid building blocks strung in long polypeptide chains, so named because of the peptide bonds that hold the amino acid links in the chains together. The precise amino acid sequence of a protein is determined by the DNA genetic code. Polypeptide chains can be several hundred amino acids long.

All but one of the amino acids (glycine) have side chains that can affect how the polypeptide will interact with water and membrane phospholipids. Two of the 20 amino acids have positively charged side chains, two have negatively charged side chains, and six have polar side chains. The charged and polar amino acids are hydrophilic. Nine amino acids are hydrophobic, having side chains that are neither charged nor polar. The amino acid sequence determines how electric charge is arranged in a protein, which determines the relationship that the protein will have with water and with the cell membrane.

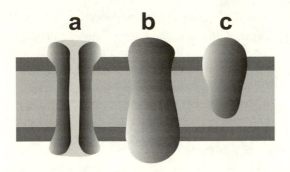

Figure 2.7. Membrane proteins.

Polypeptide chains form proteins of all shapes and sizes. One of the 20 amino acids confers a bend in the polypeptide chain, and two amino acids have side chains with binding sites that can bind two polypeptide chains together or form a closed loop within a single polypeptide chain. Some protein structures are made of a number of subunit proteins. Some membrane proteins form structures with hydrophobic middle regions and hydrophilic regions at each end, allowing them to span the membrane (Figure 2.7a, b), and some of these membrane-spanning proteins form channels through the membrane (a). There are also proteins that are confined to one side of the membrane (c). The limitless possible arrangements allow proteins to serve a vast array of functions in and around cells. Some proteins are dynamic machines with moving parts such as the membrane-spanning proteins that form the ion pumps and ion channels that produce the voltage across the cell membrane (see Chapter 3). Some proteins are dissolved in solution both inside and outside of cells, such as enzymes that perform a wide variety of functions. Some structural proteins are located inside cells to give them shape, and some are secreted forming structures outside cells, such as keratin in hair and skin and collagen throughout the body.

I am sure that most everyone beginning a study of cell biology finds it surprising that the cell membrane is like an oil slick floating on the surface of the cell. The fluid mosaic model of the cell membrane originated by Seymour J. Singer and Garth L. Nicholson in 1972 portrayed the cell membrane as a two-dimensional fluid (Singer and Nicholson 1972). Membranes exist in the world of Brownian motion. Molecules in a fluid are constantly jiggling around and jostling each other in continuous thermal motion. Because the phospholipids in the cell membrane are not tightly bound to each other, they are free to shuffle around within the plane of the membrane. Without some means of holding it in place, a phospholipid molecule at one location today could be in an entirely different location in the membrane tomorrow—hence the description of the lipid membrane as fluid. The "mosaic" part of the fluid mosaic model derives mostly from the protein structures that float in the fluid membrane and are governed by the same forces as the membrane lipids.

If you spend any time thinking about the fluid mosaic concept, sooner or later you will wonder why you do not fall apart, since there is no way to get a grip on the cell membrane. The lack of tight chemical bonds between the phospholipids and proteins in the cell membrane has enormous implications for the structure of the cell and of all living organisms. The cell membrane is not a skin that has structural strength of its own; many membrane proteins are anchored to structural proteins inside and/or outside the cell, which keeps them localized at the sites where they function. Likewise, structural connections that hold cells together or link cells to the matrix of extracellular structural proteins involve membrane-spanning proteins that bind to structural proteins on both sides of the cell membrane. In essence, the structural proteins inside the cells in your solid tissues and organs are connected to each other and/or to the structural proteins outside the cell by protein linkages that pass through the cell membranes. All complex multicellular organisms, us included, are built of interconnected scaffolds made of protein fibers produced by our cells. The protein fibers inside our cells include actin and myosin, which form the machines that produce contraction of muscle. The protein fibers outside our cells include collagen in bone, which forms a matrix for the deposition of calcium by our cells, and keratin secreted by hair follicle cells to produce hair. Our cells have built themselves quite an impressive structure in which to live, walk around, and consider life's origin and meaning.

3

THE ANIMAL BATTERY

Where does the power for nerve conduction and muscle contraction come from? In our case, mostly from the air we breathe. It wasn't until Robert Boyle (1627–1691) demonstrated that air is required for combustion that progress toward real understanding of oxidation became possible. Boyle designed vacuum pumps and showed that an animal in a sealed container quickly died when the air was pumped out (Brazier 1984). Oxidation is the principal means of releasing chemical energy from the molecules in our food, and taking in oxygen, not vital spirits, from air is the reason we breathe. Oxygen was not discovered until the 1770s.

Cells produce the energy for nerve function, muscle contraction, and everything else that our bodies do. Cells are semiautonomous machines. If a familiar metaphor helps, think of them as motor vehicles. Like cars, cells generate their own electricity from chemical fuel. For cells the fuel consists of the calorie-rich molecules obtained from food and delivered to the cells by the bloodstream. The capillaries connect to the extracellular fluid surrounding all cells. Cells take up the sugars and fats from the extracellular fluid and convert the energy they contain into a form of chemical energy that the cells can use—adenosine triphosphate, better known as ATP. Every cell in your body uses some of the energy from its supply of ATP to produce a voltage across the cell membrane called the membrane potential.

The immediate sources of power for the membrane potential are concentration gradients of ions across the cell membrane. Ionic gradients arise and are maintained by membrane proteins that serve as ion pumps, utilizing energy from ATP to transfer ions across the cell membrane. There are several ion pumps, but the sodium-potassium exchange pump is the major producer of animal electricity. The pump maintains the concentration of potassium (K^+) inside cells about 50 times higher than the concentration outside and the concentration of sodium (Na^+) outside about 10 times higher than the concentration inside. Figure 3.1 shows approximate values for the squid giant axon. The extracellular fluid consists mostly of sodium chloride, with a sodium ion concentration of about 150 millimolar (mM), while the cytoplasm inside the axon is dominated by a 150 mM potassium concentration.

Every square micrometer of the cell membrane of every cell in your body contains thousands of sodium-potassium exchange pumps. These pumps chug along continuously every second you are alive. If they stopped, you would probably have about five minutes left to live, so they matter to you very much. Each sodium-potassium exchange pump picks up three sodium ions from the cytoplasm, transfers them across the cell membrane, and releases them into the extracellular fluid. In exchange the pump picks up two potassium ions from the extracellular fluid, transfers them across the membrane, and releases them into the cyto-

	in	out
K⁺	150 mM	3 mM
Na⁺	15 mM	150 mM

Figure 3.1. Sodium and potassium concentrations inside and outside of a squid giant axon.

plasm. This exchange produces and maintains the sodium and potassium concentration gradients across the cell membrane.

Sodium and potassium also cross the membrane by flowing down their concentration gradients through membrane channels (see below), so the pump must function continuously or sodium flowing back into the cell and potassium flowing out of the cell will destroy the concentration gradients. Interrupting the supply of ATP to the pump by blocking metabolic activity with the drug ouabain causes the pump to stop functioning. The resulting flow of sodium and potassium down their concentration gradients across the membrane disrupts the osmotic balance, creating negative osmotic pressure inside the cells, causing them to take up water and swell. Without the pump, the influx of water would eventually cause the cell membrane to break open, so cells could not exist without the sodium-potassium exchange pump and other ion pumps as well. Thus, in addition to supplying electrical power to cells that transmit electrical signals, the pump is essential for maintaining the osmotic balance vital to all cells. The sodium-potassium exchange pump was unknown when most of the principles of animal electricity, including the precise mechanism of the action potential, were worked out. Jens Christian Skou discovered the pump in 1957 while an assistant professor at the University of Aarhus in Denmark. It can take time for the importance of a discovery to be recognized, and 40 years later Skou was awarded a share of the Nobel Prize in Chemistry for his discovery.

The sodium-potassium exchange pump forms an ion channel through the cell membrane (see Nicholls et al. 2001). The polypeptide chains lining the channel contain negatively charged amino acids so that only positively charged ions can pass through, and the channel is

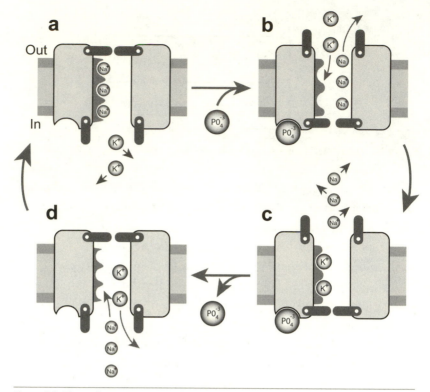

Figure 3.2. The sodium-potassium exchange pump cycle.

of a size that admits only the small ions sodium and potassium. Like many different kinds of ion channels, the sodium-potassium exchange pump protein is equipped with gates that open and close. The pump has two: one that operates at the entrance to the channel inside the cell and the other at the entrance outside the cell (Figure 3.2). Proteins that function as molecular machines that have moving parts such as channel gates operate by altering the three-dimensional configuration of their polypeptide chains, known as conformational change. The sodium-potassium exchange pump has just two conformations: either the outer gate is closed and the inner gate is open (a, d), or the outer gate is open and the inner gate is closed (b, c). It is important to the functioning of the pump that both gates are never open simultaneously. The status of the gates is not the only difference between the two conformations. When the internal gate is open the charged amino acids lining the channel are configured to bind to three sodium ions (a), and when the external gate is open, the charges in the channel are configured to bind to two potassium ions (c). Thus, sodium ions bound inside the channel always come from inside the cell (d), and potassium ions bound inside the channel always come from outside the cell (b).

There is an ATP binding site exposed on the cytoplasmic side of the pump protein. ATP briefly binds to the site and transfers a high-energy phosphate (PO_4^{-3}) to the pump protein in a process known as phosphorylation. Phosphorylation supplies the energy to change the pump from the conformation with the inner gate open (Figure 3.2a) to the conformation with the outer gate open (b). When the pump reverts to the conformation with the inner gate open, the phosphate dissociates from the channel (c–d). There is normally a continuous supply of ATP inside the cell, so the pump continuously flips back and forth between the two conformations. However, the pump cannot change conformation without a trigger. The binding of sodium in the channel triggers phosphorylation of the channel and the change to the conformation with the outer gate open, and the binding of potassium inside the channel triggers dephosphorylation and the change back to the conformation with the inner gate open.

This all sounds rather complex, but once the properties of the channel are put together it is easy to see how the pump works. In the dephosphorylated conformation, only the inner gate is open, and three sodium ions from inside the cell bind inside the channel (Figure 3.2a). Sodium binding triggers the change to the phosphorylated conformation in which only the outer gate is open (b), the binding sites release the three sodium ions outside the cell in exchange for two potassium ions (b–c). To complete the cycle, potassium binding triggers the return to the dephosphorylated conformation with only the inner gates open, and the binding sites release the two potassium ions inside the cell in exchange for three sodium ions (d–a). The net result is that three sodium ions have been transferred outside the cell (c) in exchange for two potassium ions entering the cell (a).

The cycle continues over and over again as long there is sodium inside the cell to be removed, potassium outside the cell to be gathered, and the supply of ATP holds out. The exchange ratio of three sodium ions for two potassium ions seems reasonable since potassium ions are larger than sodium ions and potassium and sodium have to occupy the same space within the channel. The three-to-two exchange of electric charge has a small direct effect on the membrane potential, but the slight negative effect on the membrane potential of ejecting one more positive ion than is taken up during each cycle has little, if any, mechanistic consequence. While the sodium-potassium exchange pump is the ion pump most relevant to electricity production by cells, there are others also of vital importance; for example, chloride pumps maintain the chloride ion gradient across the cell membrane.

A word is in order about the diagrams I use to depict the sodium-potassium exchange pump, as well as the voltage-gated potassium channel and the voltage-gated sodium channel, which are presented further along in this chapter. The real gates look nothing like pinball flippers. Gates are part of the protein structure of channels and they vary in their operation. Some actually are hinged lids that close the channel by covering the opening, but others operate as collars that close the channel by causing a constriction that prevents ions from passing through. Regardless of the details, all gates operate by changes in the conformation of the protein. I have constructed the pump and channel diagrams to represent functional reality as much as possible. In Figure 3.2, sodium and potassium ions are represented as spheres, which they are, and potassium ions are depicted as about 20% larger in diameter than sodium ions, which they are. The configurations of the ion-binding sites are depicted as three concavities that fit the sodium ions or two concavities that fit the potassium ions. The concavities are intended to simply convey the selective affinity for each species of ion and not the actual shape of the protein region to which the ions bind. The real binding sites are actually oxygen atoms in the protein structure, each of which possesses a partial negative charge. Because of the strength of the electric force, these negative charges must always be neutralized, so when the conformation of the pump protein changes, the binding sites do not release their sodium or potassium and then wait for the other ion to flow in and bind as might be supposed when viewing Figure 3.2b and d. In this case artistic license was necessary to depict the exchange of ions. Artistic license was also operating when the binding site was depicted on only one side of the channel. I emphasize these points partly to caution readers who might wish to pursue some of the topics I present here in other books or on the Web. Representations

Figure 3.3. Molecular model of the structure of a voltage-gated potassium channel. Five potassium ions are shown, four of which bind to negatively charged amino acids lining the channel. (Morais-Cabral, Zhou, and MacKinnon 2001; adapted with permission from Macmillan Publishers Ltd., *Nature* 414:37–42, copyright 2001.)

of complex proteins such as pumps and channels are diverse and rarely look like the structures they represent. If the point of the representation is to show something of polypeptide structure, membrane proteins are often depicted as computer-generated diagrams such as the potassium channel depicted in Figure 3.3. Such diagrams capture some aspects of the real structure, but by no means capture the precise appearance of the protein. Readers can find many different images of voltage-gated potassium channels on the Web. For the purposes of communicating something of how the membrane pumps and channels work I prefer my pinball-flipper diagrams, because trying to portray the structural details of pumps and channels gets too complicated and tends to enter into the nebulous area of current theory and research.

Why don't the thousands of sodium-potassium exchange pumps in each cell, pumping day and night, eventually pump all the sodium out of the cells and collect all the potassium from the extracellular fluid into the cells? First of all, the fluid in the extracellular space surrounding all of our cells is connected to the bloodstream via the capillaries. The ionic composition of blood and extracellular fluid is mostly sodium and chloride with lesser amounts of calcium and other ions. The kidneys tightly regulate this ionic composition. For example, if there is an excess of sodium in the blood, the kidneys increase their excretion of sodium, and if there is a shortage of sodium in the blood the kidneys slow their excretion, allowing sodium in the diet to restore the proper physiological level. The kidneys work the same for potassium and other ions. The sodium pumped out of the cells cannot under normal circumstances increase the sodium in the extracellular space beyond its physiological limit, and the potassium taken up by the cells cannot deplete the potassium below its physiological limit. Also, the sodium-potassium exchange pump cannot remove all the sodium from the cell or increase the potassium inside to extremely high levels because there is always flow of sodium into the cell and potassium out of the cell driven through membrane channels by their concentration gradients (see below). Rather, there is a balance between the pumping of ions and the flow of ions down their concentration gradients that keeps sodium and potassium concentrations and the concentrations of other ions inside the cells at precise levels.

THE POTASSIUM BATTERY

All that potassium inside the cells is in constant thermal motion, banging on the membrane, trying to get out. When a potassium ion bangs against the phospholipids in the cell membrane it is repelled by the hydrophobic lipids in the membrane core, so its escape attempt by that route is useless. However, there are proteins that span the cell membrane, forming channels through which ions can pass. The most important channels in generating electricity are the voltage-gated potassium channels and voltage-gated sodium channels. They are equipped with molecular gates that can open and shut in response to changes in the membrane potential. When a potassium ion strikes the membrane at a potassium channel with an open gate, it is confronted with an opening into which it will fit lined with negatively charged amino acids (Figures 3.3 and 3.4), which can neutralize its positive charge, allowing the potassium ion to shed its hydration shell and pass through the channel. (Note that the potassium channel is made up of four protein subunits surrounding the hole through the membrane, but the subunit nearest the reader has been omitted from Figure 3.3 to better reveal the potassium ions, and only two subunits are shown in Figure 3.4.)

Positive ions such as calcium that are too big do not fit in the potassium channel, and sodium ions, which are smaller than potassium ions, cannot interact with the negative charges inside the channel as strongly as potassium and so are excluded by competition with potassium. Negative ions such as chloride have no chance of entering the potassium

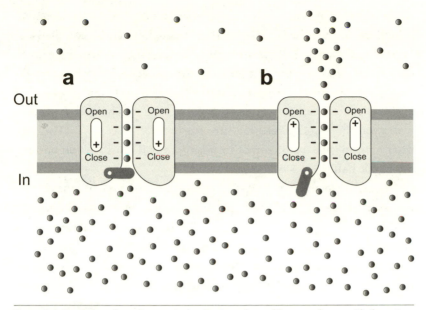

Figure 3.4. Depiction of a voltage-gated potassium channel in a membrane with the gate closed (a) and open (b).

channel regardless of their size because of the repulsion of the fixed negative charges lining the channel for any negative ion.

When I first learned about membrane potentials in the 1960s, nobody knew anything about the structure of ion channels. I thought of them as holes through the membrane, but they are nothing like that; they are like short electric wires connecting the inside and outside of the cell. Because of the immense strength of the electric force, the negative charges lining the potassium channel must always have positive ions nearby, so potassium channels are always filled with four potassium ions lined up single file (see Köpfer et al. 2014). Thus, the potassium ions behave like mobile charges that can move as a group in relation to the fixed negative charges lining the channel. You can see that potassium channels behave like electrical wires, with fixed charges of one sign, in this case negative charges on amino acids lining the channel, and mobile charges of opposite sign, in this case the potassium ions inside the channel. In effect, potassium channels are protein wires that conduct potassium ions according to the same principles as apply to metal wires that conduct electrons.

When a potassium ion strikes an open potassium channel from inside the cell with sufficient energy, it breaks out of its hydration shell and knocks the lineup of four potassium ions along the channel, binding to the negative charge at the first position. Since there is only room for four potassium ions lined up single file in the channel, the potassium ion that had occupied the last position is knocked out of the channel, acquires a hydration shell, and is released into the extracellular fluid. (Think of what happens when a lineup of four billiard balls is struck on one end by the cue ball.)

The frequency at which potassium ions strike the membrane at open potassium channels with enough energy to enter will depend upon the concentration of potassium ions, the density of open potassium channels in the membrane, and the average kinetic energy of the potassium ions, that is, the temperature. Because the density of channels and the temperature are always the same inside and outside the cell, all that matters is the concentration difference. If the 50:1 potassium concentration gradient was the only force operating, there would be an average of 50 potassium ions that enter the channel from the inside of the cell for every potassium ion entering from the outside. Since a potassium ion entering the channel always bumps a potassium ion out the other end, the concentration gradient would produce a continuous flow of potassium ions out of the cell until the potassium ions were depleted to a concentration inside the cell matching the concentration outside. This all arises from the simple principle of diffusion. Molecules in solution move around in all directions in perpetual motion, causing them to spread out from regions of higher concentration to regions of lower concentration. (Place a drop of ink in a container of water and watch it happen.)

In 1902 German physiologist Julius Bernstein (1839–1917) proposed the first quantitative theory of animal electricity (Hodgkin 1963). Bernstein's theory explains why potassium does not continuously flow out of cells down its concentration gradient. The reason—you guessed it—is the immense electric force. While potassium ions cannot stand to be too near one another they also hate to be alone. Like all ions they cannot get too far away from ions of opposite sign that neutralize their charge. There are enormous electrical consequences of moving even a small number of potassium ions from the inside to the outside of the cell unaccompanied by negative ions that neutralize their charge. When a potassium ion moving down its concentration gradient leaves the cell, it carries a positive charge into the extracellular solution and leaves behind a negative charge inside the cell. The accumulation of excess positive charge outside the cell and excess negative charge inside the cell produces a voltage difference across the cell membrane—the membrane potential. Since it is the charge difference that matters, not the absolute number of charges, it is the convention to set the voltage of the extracellular fluid at zero, and the membrane potential then is the voltage inside the cell. Thus, the flow of potassium down its concentration gradient out of the cell produces a negative membrane potential.

As soon as the membrane potential develops, the concentration gradient ceases to be the only force driving potassium across the membrane. The negative voltage produced inside the cell by the outflow of potassium produces an electric field that tends to draw potassium back in—the opposite direction of the flow of potassium driven by the concentration gradient. At some voltage the membrane potential reaches a magnitude where it would draw just as many potassium ions back into the cell as the concentration gradient drives out of the cell. At this voltage there will be no further net transfer of charge across the cell membrane. This is an equilibrium, where the membrane potential doesn't change because there is no longer net potassium current into or out of the cell, and the voltage at which this occurs is called the potassium equilibrium potential.

Bernstein calculated the potassium equilibrium potential by using the Nernst equation developed by physical chemist Walter Nernst (1864–1941) about 15 years earlier (see Nicholls et al. 2001; Kandel et al. 2013). The potassium equilibrium potential is calculated from the ratio of the concentrations of potassium outside and inside the cell (Figure 3.5; note that brackets designate that the concentration of the enclosed ion is used in the calculation). (Interested readers, see the detailed explanation of the Nernst equation in the endnote.[1] Readers wishing to ignore the Nernst equation, forge on; be assured that your understanding of the subject will not be significantly impaired.)

Calculated using the Nernst equation, the concentration gradient produced by a 50-times-greater potassium concentration inside the cell than outside yields a voltage of about −99 millivolts (mV) at a room temperature of 20°C. This may not seem like much, but a millivolt is

$$V_K = \frac{RT}{F} \ln\left(\frac{[\text{K}^+]_{\text{out}}}{[\text{K}^+]_{\text{in}}}\right)$$

Figure 3.5. The Nernst equation for the potassium equilibrium potential.

one-thousandth of a volt, so 100 mV is one-tenth of a volt. The potassium batteries of just 15 cells connected in a series would produce a voltage of nearly 1.5 volts—the same voltage as an AA battery—not bad for 15 tiny cells. Because of the tremendous strength of the electric force, the transfer of potassium ions out of the cell that produces a voltage of −99 mV is very small—much too small to make a measurable dent in the 150 mM internal potassium concentration.

The potassium ions driven back into the cell by the negative membrane potential travel through the same potassium channels as the potassium ions driven out of the cell by the concentration gradient. This is best pictured as two separate and opposed processes operating on a single potassium channel. The voltage across the membrane exerts a force on the lineup of four mobile potassium ions in the channel, just like a voltage from a human-made battery exerts a force on the mobile electrons in a metal wire. Of course, potassium cannot move through the channel into the cell unless potassium ions enter the channel from the outside and leave the channel from the inside. The constant force of the electric field nudging the four potassium ions in the channel toward the inside of the cell reduces the energy required for potassium ions to enter the channel from outside the cell and increases the force required for potassium ions to enter the channel from inside the cell. At the potassium equilibrium potential produced by a 50:1 concentration gradient, an individual potassium ion striking the channel entrance from outside the cell will have a 50 times greater probability of entering the channel than a potassium ion striking the channel entrance from inside the cell. Thus, the equilibrium potential is the voltage where the 50-times-greater frequency of potassium ions striking the channel opening on the inside of the cell is exactly compensated by a 50-times-greater probability that any individual potassium ion striking the channel from the outside will have sufficient energy to enter the channel.

This all may seem very reasonable, but other ions are present and there are other ion channels in the membrane that they can pass through, so the membrane potential is not determined by the potassium concentration gradient alone. The two other major contributors to the membrane potential that we need to consider are sodium and chloride. The concentrations of potassium, sodium, and chloride inside and outside the cell and their individual equilibrium potentials calculated from the Nernst equation are given in Figure 3.6. All three ions have very different equilibrium potentials, but the membrane can only have one membrane potential, so something has to give.

It turns out that the contribution of each ion to the membrane potential is weighted by the permeability of the membrane to that ion relative to the permeability to all the other ions; the more permeable the membrane is to an ion, the greater the relative contribution of that ion to the membrane potential. The key to understanding how this works is realizing that when the currents of all the participating ions are summed up, the resulting net ionic current across the membrane must be zero for the membrane potential to be stable; all the outward ionic currents together must equal all the inward ionic currents. Any difference between the outward and inward ionic currents amounts to a net current across the membrane capacitance, which changes the membrane potential. As explained for potassium in detail above, an ion contributes no net current when the membrane potential is at its equilibrium potential. The current supplied by any given ion progressively increases the farther the membrane potential deviates from the equilibrium potential of the ion. A highly permeant ion carries a relatively large current if the membrane is far from its equilibrium potential, whereas a less permeant ion carries a smaller current. To achieve no net ionic current across the membrane, the membrane potential must settle at a value that is closer to the equilibrium potentials of highly permeant ions, in order to reduce the current carried by those ions to a low level where it is compensated by the current carried by the less permeant ions. Basically, it is a trade-off between the electrical driving force and membrane permeability. Simply stated, at a stable membrane potential highly permeant ions must be subject to a relatively low electrical driving force to keep their current

	in	out	V_{ion}
K⁺	150 mM	3 mM	-99 mV
Na⁺	15 mM	150 mM	+58 mV
Cl⁻	9 mM	125 mM	-66 mV

Figure 3.6. Monovalent ion concentrations inside and outside a squid giant axon and the equilibrium potential for each ion at 20°C (293 K).

relatively low, while less permeant ions must be subject to a relatively high electrical driving force to bring their current high enough to compensate.

Alan Hodgkin and Andrew Huxley, who worked out the mechanism of conduction of the action potential by investigating squid giant axons (see Chapter 4), found that at rest, when there is no action potential, the squid axon membrane is about 25 times more permeable to potassium than to sodium and about 11 times more permeable to chloride than to sodium. Therefore, the potassium and chloride equilibrium potentials dominate the resting potential, and the sodium equilibrium potential has relatively little influence. The membrane potential produced by multiple ions is calculated from the Goldman-Hodgkin-Katz equation (Figure 3.7), which is an expanded version of the Nernst equation. (Readers who ignored the Nernst equation, ignore the Goldman-Hodgkin-Katz equation and forge on!) The membrane potential is calculated from the sum of the concentrations of the ions that drive inward current, each weighted by their relative permeability, divided by the sum of the concentrations of the ions that drive outward current, also weighted by their relative permeability. So the Goldman-Hodgkin-Katz equation is just a way of squeezing multiple ions into the Nernst equation. Note that the chloride concentrations are flipped, with the concentration inside the cell in the numerator and the concentration outside the cell in the denominator, because by convention current is treated as flowing in the opposite direction of the flow of a negative ion.

$$V_m = \frac{RT}{F} \ln\left(\frac{p_K[K^+]_{out} + p_{Na}[Na^+]_{out} + p_{Cl}[Cl^-]_{in}}{p_K[K^+]_{in} + p_{Na}[Na^+]_{in} + p_{Cl}[Cl^-]_{out}}\right)$$

Figure 3.7. The Goldman-Hodgkin-Katz equation.

The relative permeability of the resting membrane of the squid giant axon to potassium, chloride, and sodium determined by Hodgkin and Huxley yield a resting potential of about −70 mV.

THE ELECTRIC MODEL OF THE CELL MEMBRANE

When a current in the form of potassium ions is conducted through potassium channels in the cell membrane, it passes through a resistance and a change in voltage. In circuit diagrams, potassium channels are represented by a battery with a voltage equal to the potassium equilibrium potential connected in series with a resistor representing the resistance of the channel to potassium current as shown in Figure 3.8a. The circuit representations of sodium channels and chloride channels are shown in b and c.

Talking about ionic resistances can get confusing because an increase in the permeability for an ion decreases the resistance. We will speak of ionic conductance (g) instead. Conductance is just the reciprocal of resistance: $g = 1/R$. Recall from Chapter 1 that the familiar form of Ohm's law is $I = V/R$, where current (I) is equal to the voltage (V) driving the current divided by the resistance (R) of the circuit to the flow of current (see Figure 1.1). Substituting conductance for resistance, Ohm's law becomes $I = Vg$, where current equals the voltage driving the current times the conductance of the circuit. Since doubling the voltage or doubling the conductance both double the current, it is much easier to talk about conductance than resistance.

Another potential source of confusion is the relationship between conductance and permeability. At first sight it might seem that ionic permeability and ionic conductance should be directly proportional so that, for example, increasing the sodium permeability 45-fold as happens during the action potential in squid axon should increase the sodium

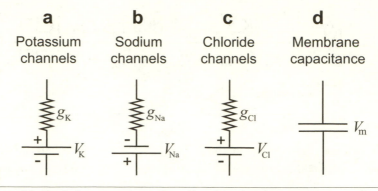

a	**b**	**c**	**d**
Potassium channels	Sodium channels	Chloride channels	Membrane capacitance

Figure 3.8. Circuit diagram representations of ion channels and the membrane capacitance.

conductance of the membrane 45-fold. But this is not the case because the permeability only depends upon the density of open ion channels in the membrane, whereas the conductance depends upon the density of open ion channels and the availability of ions to go through them. If there were no sodium ions inside or outside the cell, it would not matter how many open sodium channels were in the membrane, the conductance of the membrane to sodium would be zero. Therefore, the relative permeability of the membrane to potassium, sodium, and chloride is not the same as the relative conductance of the membrane to these ions; relative conductance cannot be used in place of relative permeability in the Goldman-Hodgkin-Katz equation. Conductance of the membrane is determined by direct electrical measurement (see Chapter 4).

Although ions can pass through membrane channels, phospholipids make up most of the membrane surface area in most regions of the cell so overall the membrane has a much higher electrical resistance than the ionic solutions inside and outside the cell. Therefore, as previously discussed, the membrane can be viewed as the insulating layer in a capacitor between the ionic solutions inside and outside the cell, which form the conductive plates (see Chapter 1). The membrane potential is represented in circuits as the voltage across a capacitor (Figure 3.8d).

A circuit model of the cell membrane can be constructed with these electrical representations of ion channels and the membrane capacitance. First consider a circuit with only potassium conductance across the membrane. In order to see the impact of potassium conductance, a switch is installed in the circuit diagrams in Figure 3.9 that can turn the potassium conductance on and off. The circuit is much like the circuit depicted in Chapter 1 (Figure 1.5), but with an important modification. Voltage is always relative, and let's set the voltage outside the cell to zero so that the voltage inside the cell is the membrane potential. Figure 3.9a depicts the circuit before the switch is turned on, and the voltages across the potassium channel conductance and the membrane capacitance are accordingly also zero. When the switch is turned on, current flows clockwise around the circuit, resulting in an outward current through the potassium conductance and an inward current across the membrane capacitance. Figure 3.9b shows the circuit at a time after the switch has been turned on when the membrane capacitance is charged to half the value of the potassium equilibrium potential. Therefore, there is a voltage drop of -49.5 mV as the inward current crosses the membrane capacitance. Next the current crosses the potassium equilibrium potential, which raises the voltage by $+99$ mV, leaving $+49.5$ mV driving outward current across the potassium conductance. This makes sense. Simply stated, the voltage driving the potassium current is the potassium equilibrium potential minus the voltage produced by the charge that has built up on the membrane capacitance; that is, minus the membrane potential. Once the charge on the membrane capacitance reaches the potassium equilibrium potential (Figure 3.9c), there is no voltage remaining to drive current across the potassium conductance, so the current stops, and the membrane potential remains steady at the potassium equilibrium potential.

This analysis reveals two important features of current flow across cell membranes. The first emerges directly from the application of Ohm's law to membrane circuits. The voltage driving the current carried by any particular ion across the membrane is not just the membrane potential. Rather, it is the difference between the membrane potential and the equilibrium potential for that ion. For the potassium current Ohm's law becomes $I_K = (V_m - V_K)g_K$. Thus, when the membrane is at the potassium equilibrium potential, the voltage driving the potassium current is zero, so there is no net potassium current crossing the membrane. The same relationship holds for sodium, chloride, and any ion that can cross the membrane (Figure 3.10).

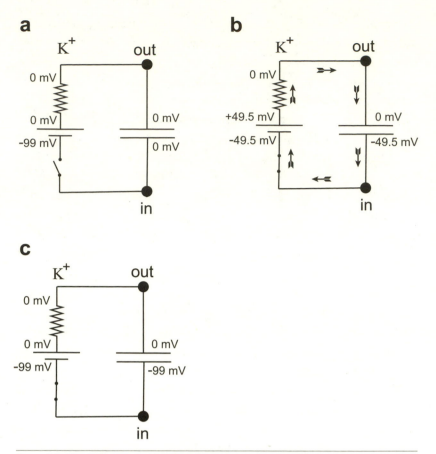

Figure 3.9. Circuit diagram model of a cell membrane showing only the potassium conductance and the membrane potential, and with a switch installed.

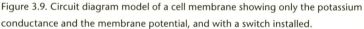

The second principle of current flow across membranes revealed by the circuit analysis is that the total current across the membrane must always be zero. At first this may seem surprising, but it is easy to see why when examining the current flow in the circuit in Figure 3.9. When the switch is turned on, current supplied by the potassium battery flows out of the positive terminal, and the same amount of current must always flow into the negative terminal. You might ask what about the charge that builds up on the membrane capacitance? The buildup of charge does not represent a blockage of the current, since the charge drives current across the capacitor in the form of an electric field. (Remember from Chapter 1 that current can flow in the form of electrons, ions, or an elec-

Ohm's Law: $I = Vg$

$$I_K = (V_m - V_K)g_K$$

$$I_{Na} = (V_m - V_{Na})g_{Na}$$

$$I_{Cl} = (V_m - V_{Cl})g_{Cl}$$

$$I_{ion} = (V_m - V_{ion})g_{ion}$$

Figure 3.10. Ohm's law applied to ionic current through membrane channels.

tric field.) It is a basic principle of electricity that current in circuits always flows in closed loops, and when current is flowing all the current entering the loop must return; no current gets stalled along the way. Therefore, in the circuit in Figure 3.9b, the ionic current leaving the cell through the potassium channels exactly equals the capacitative current entering the cell across the membrane capacitor.

Now let's make the circuit diagram more realistically represent the membrane by including sodium channels and chloride channels. The circuit diagram in Figure 3.11 depicts the membrane at a resting potential of −70 mV. The voltage across each of the circuit elements is indicated. Starting clockwise from outside the cell, the voltage drops from 0 to −70 mV across the membrane capacitance. This is the resting potential, which is stable because no current is flowing through the membrane capacitance. Carrying on in the clockwise direction, the potassium, sodium, and chloride channels provide three parallel paths back across the membrane. The total voltage change across each path is the same since the potential must go from −70 mV inside the cell back to zero outside the cell regardless of the path taken. However, since the three ions have different equilibrium potentials, the +70 mV change is distributed differently between the equilibrium potential and the conductance of each ion.

Considering each ion in turn, the path out through the potassium channels first encounters a +99 mV increase upon crossing the potassium equilibrium potential, which results in a voltage of +29 mV that

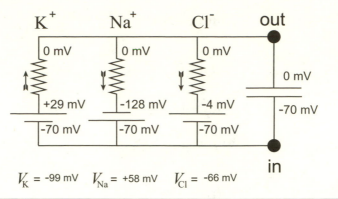

Figure 3.11. Circuit model of the resting potential with voltages across each of the circuit elements indicated.

must drop to zero upon crossing the potassium conductance to reach the outside of the cell. Thus, there is a voltage of +29 mV driving outward potassium current (arrow) across the potassium conductance. By a similar analysis, the path through the sodium channels encounters a voltage drop of −58 mV when crossing the sodium equilibrium potential, which adds to the −70 mV resting potential, resulting in a voltage of −128 mV across the sodium conductance driving inward sodium current through the sodium channels. Lastly, there is a rise in voltage of +66 mV encountered upon crossing the chloride equilibrium potential resulting in a voltage of −4 mV driving inward chloride current through the chloride channels. (Recall that current is defined as moving in the direction of positive charge, so −4 mV drives chloride current into the axon, the opposite direction of the actual movement of the chloride ions carrying the current.)

At the resting potential of −70 mV there is no current flowing across the membrane capacitance, so the net ionic current across the membrane must also be zero. Therefore, at the resting potential, the outward potassium current must be equal to and opposite of the inward sodium and chloride currents. Chloride contributes little current since the membrane potential is only −4 mV away from the chloride equilibrium potential. Thus, at the resting potential the major ion flows driven by the concentration gradients consist of a nearly equal exchange of sodium flowing into the cell and potassium leaving the cell. At the resting potential the potassium and sodium currents are equal and opposite because the membrane potential is nearer to the potassium equilibrium potential by just the right magnitude so that the greater permeability of the membrane for potassium is compensated by the greater voltage driving the sodium current. Because of the ongoing exchange of sodium for potassium, the resting potential is not a true equilibrium, and the potassium and sodium gradients would dissipate were they not continuously maintained by the sodium-potassium exchange pump.

You may find it curious that ion channels are represented by a battery and resistance connected in series in circuit diagrams and wonder what can the voltages of +29 mV "inside" the potassium channel, −128 mV "inside" the sodium channel, and −4 mV "inside" the chloride channel possibly mean. Those voltages do not really exist inside the membrane because equilibrium potentials are not actually voltages that can be measured directly. Taking potassium as an example, the potassium equilibrium potential just represents the impact of the potassium concentration gradient on the flow of potassium across the membrane in electrical terms; that is, the potassium concentration gradient produces a flow of potassium across the membrane equivalent to the potassium current that would be produced by a battery with a voltage of −99 mV. Therefore, the potassium concentration gradient can be represented in a circuit diagram by a −99 mV battery. Since potassium current flowing outward through potassium channels is always impacted by both the potassium equilibrium potential and the potassium conductance when crossing the membrane, they are connected in series in the circuit representation of the potassium channel. Although connected in series, in the real world both have to be crossed at once when the membrane is crossed; there is no stopping in between. Thus, the circuit representations of the ion channels do a nice job of modeling the electrical behavior of the membrane that arises from the electrical consequences of ionic concentration gradients operating according to Ohm's law.

VOLTAGE-ACTIVATED GATING

The action potential was once regarded as the greatest mystery of how the nervous system works. The discovery of the mechanism of the action potential by Hodgkin and Huxley and how the action potential travels along the axon are the subjects of Chapters 4 and 5. We will set the stage

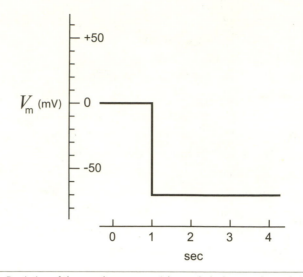

Figure 3.12. Depiction of the membrane potential recorded when an electrode is inserted into an axon at the one-second mark.

with a description of how the conductance changes of the sodium and potassium channels produce the voltage changes across the membrane during the action potential. While each individual ion channel is an on-off switch, collectively ion channels do not function in a simple on-off manner. Collectively, they function more like dimmer switches (or water faucets) because the control is graded. The control mechanisms are different for channels with different functions. Methods to directly measure the membrane potential by inserting an electrode directly into a cell or axon were developed in the mid-twentieth century. How this was accomplished is discussed in Chapters 4, 5, and 6. For now, imagine the voltage is displayed on a cathode-ray oscilloscope screen or computer screen. The cathode-ray beam or computer is tracing out the recorded voltage over time from left to right as indicated by the timescale in Figure 3.12. The trace starts at 0 mV and progresses across the screen from left to right. An electrode is inserted into an axon at the 1-second mark, and the recorded voltage abruptly drops from 0 to −70 mV, the resting potential, and is maintained for the duration of the trace. (This is actually a "thought" experiment since the vast majority of axons are not large enough in diameter to tolerate an electrode being inserted across their membrane—more about that in Chapter 4.)

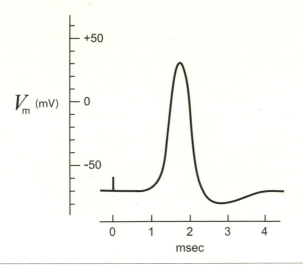

Figure 3.13. Depiction of an action potential recorded as it passes by an intracellular electrode.

Now with the electrode inside the axon, the speed at which the trace crosses the screen is increased 1,000 times by turning a knob on the oscilloscope or computer so that changes in voltage can be observed on a millisecond (msec) timescale (Figure 3.13). Imagine there is a stimulating electrode located on the axon somewhere upstream of the recording electrode. A shock is delivered to the axon through the stimulating electrode at time mark zero, which is immediately detected as a brief deflection of the voltage, called the stimulus artifact. It is produced at the stimulating electrodes and is conducted through the external solution at near light speed and picked up by the recording electrode, providing a convenient indication of the time the action potential is triggered by the stimulating electrode. About 1 msec after the stimulus artifact, the membrane potential shoots up 100 mV to a peak of +30 mV, then drops back down, slightly undershooting the resting potential to about −80 mV, and finally recovers to the resting potential value of −70 mV. The rise to peak takes just under 1 msec, and the action potential from start to finish lasts about 3 msec. This is the time course of the action potential as it would be recorded passing by a recording electrode. Action potentials vary in their exact parameters. These parameters, which fall within the normal range, have been selected to yield convenient numbers.

There is an important but potentially confusing convention used when describing the action potential. Since the resting potential is a difference in electric charge, the resting membrane is described as "polarized," and a change in the membrane potential in the positive direction toward zero is described as a "depolarization" because the difference in electric charge is decreased. This makes sense, but somehow it has become customary to refer to the entire action potential as a depolarization, which strictly speaking is not correct because after the rise of the action potential passes through zero the membrane potential is positive, so it is actually polarized in the opposite direction. I know of no better term for the rise of the action potential, so I follow the convention and refer to the entire rise of the action potential as depolarization.

Hodgkin and Huxley found that a large increase in the permeability of the membrane to sodium produces the rise of the action potential. In the squid axon membrane, the permeability to sodium increases 45-fold from a value that is 1/25th the permeability to potassium to a value that is 20 times the permeability to potassium in less than a millisecond (Hodgkin 1992). This tremendous increase in the relative permeability to sodium causes the membrane potential to shoot upward toward the sodium equilibrium potential. However, the membrane potential only reaches +30 mV because the permeability to sodium does not last—it automatically shuts itself off. In addition, the permeability to potassium doubles, and both of these changes bring the membrane potential back down toward the potassium equilibrium potential so that the entire action potential is over in just a few milliseconds. These permeability changes are produced by the opening and closing of voltage-activated gates on the sodium and potassium channels. Chloride channels have no gates and so are always open. While chloride influences the voltage achieved throughout the duration of the action potential, it plays no mechanistic role in generating the action potential. Therefore the discussion will focus on the sodium and potassium channels.

The basic scenario of the membrane conductance changes and currents associated with the action potential was worked out long before anyone knew anything about the structure of ion channels in membranes. The voltage-gated sodium channels that drive the action potential are formed by four protein subunits surrounding a hole through the membrane; only two subunits appear in the two-dimensional depiction

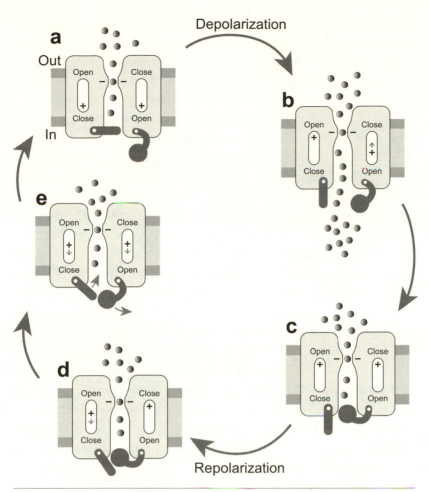

Figure 3.14. The cycle of activation and inactivation of the voltage-gated sodium channel.

of a sodium channel in Figure 3.14. The hole formed by the sodium channel proteins is not a narrow tube like the hole formed by potassium channel proteins. Instead, it has an hourglass profile with larger openings on both the outside and inside of the cell membrane, both of which funnel down to a narrower constriction in the middle—big enough for a sodium ion along with part of its hydration shell to pass through. The constriction accommodates just a single sodium ion, unlike the potassium channel that accommodates a lineup of four potassium ions.

Voltage-gated sodium channels have two gates located at the inside entrance to the channel: the sodium activation gate depicted as a pinball flipper mounted on the left subunit of the channel protein and the sodium inactivation gate depicted as a ball on a stalk mounted on the right subunit in Figure 3.14. The membrane potential regulates the conformational changes that open and close these gates. At the resting potential sodium activation gates favor the closed position and are opened by depolarization, whereas sodium inactivation gates favor the open position at the resting potential and are closed by depolarization.

The mechanism of opening and closing the sodium activation gate is known in some detail. The polypeptide chain of a sodium channel protein subunit contains a pair of alpha helices that pass through the membrane side by side. An alpha helix is a helical winding of the polypeptide chain that forms a rigid rod in the structure of a protein. One of these alpha helices has a lineup of positively charged amino acids on its surface, which faces and interacts with a lineup of negatively charged amino acids exposed on the surface of the other alpha helix. The positive and negative charges on the two alpha helices neutralize each other. The positively charged alpha helix is linked to a mobile region of the sodium channel protein that forms the sodium activation gate, while the negatively charged alpha helix is fixed in position. The positively charged alpha helix is also longer than the negatively charged alpha helix so that its positive charges extend into the extracellular fluid and the cytoplasm where their positive charges interact with the polar water molecules, much like positive ions in solution. The positively charged alpha helix moves through the channel protein in response to changes in the membrane potential. The negative charge of the resting potential pulls the positive rod toward the inside of the cell tending to close the sodium activation gate, and depolarization moves the rod toward the outside of the cell, tending to open the sodium activation gate. Thus, the positively charged alpha helix rod can be thought of as a handle that opens and closes the sodium activation gate in response to changes in membrane potential.

Different gates vary in their molecular details, but all voltage-activated gates must, in principle, involve the movement of mobile charges that change the conformation of the channel protein in response to changes in membrane potential. Like the sodium activation gates, the sodium inactivation gates and the potassium activation gates are also known to be associated with positively charged segments of polypeptide chains that move in response to changes in membrane potential. In Figure 3.4 depicting the potassium channel and Figure 3.14 depicting the sodium channel, I have represented all voltage-activated gates as controlled by positively charged switches that move inward when the membrane potential is polarized and outward when the membrane is depolarized.

Channel gates function within the world of Brownian motion. While channel gates are not free to move around in solution like dissolved molecules, they are subject to their own thermal motion and the thermal energy of molecules bumping into them, so they tend to open and close apparently spontaneously. The effect of the membrane potential is to bias the population of gates toward the open or closed conformations and so determine what fraction of the gates are, on average, switched open versus closed. For example, at the resting potential of −70 mV, potassium activation gates are spontaneously flipping open and closed in thermal motion, and at any given moment about 50% of the potassium channel gates are open. Gating mechanisms vary in three important ways: the direction of the voltage dependence, the membrane potential at the end points where the gates are either virtually all open or closed, and the kinetics, that is, how fast the gates respond to changes in the membrane potential.

At the resting potential of −70 mV the vast majority of the sodium channels have their activation gates in the closed position and their inactivation gates in the open position (Figure 3.14a). Since both sodium channel gates must be open for sodium to pass through, the permeability of the resting membrane to sodium is very low. To initiate the action potential something has to start the process of opening more sodium activation gates by depolarizing the membrane from rest to the value of the membrane potential called the threshold. The initial depolarization that triggers the action potential can be provided in several ways: the binding of a neurotransmitter to receptors on the neuron; the opening of specialized membrane channels such as in heart pacemaker cells; or current injected by a stimulating electrode during an experiment. Once the action potential has been initiated, the region of axon at the leading edge of the action potential must be continuously depolarized to threshold for the action potential to travel along the axon. The rising phase of the action potential is divided into two phases: the passive rising phase from the resting potential to the threshold and the active rising phase from threshold to the peak. Some of the electric current that enters the axon through sodium channels during the active rising phase extends forward within the axon to depolarize the membrane ahead producing the passive rising phase, which maintains the forward travel of the action potential. The mechanisms that produce the passive rising phase of the action potential are discussed in detail in Chapters 4 and 5.

Sodium activation gates respond quickly to changes in membrane potential. They have fast kinetics—think of them as well oiled. So the initial depolarization to threshold quickly opens some of the sodium activation gates (Figure 3.14b). Since sodium current entering the cell causes the membrane potential to depolarize, and since depolarization opens more sodium activation gates, a positive feedback cycle is set up: depolarization opens some of the sodium activation gates, which increases the sodium conductance, producing inward sodium current, which further depolarizes the membrane potential, opening more sodium activation gates—over and over again. This positive feedback cycle during the rising phase of the action potential tremendously increases the conductance of the membrane to sodium, producing inward sodium current that drives the membrane potential toward the sodium equilibrium potential. During the rise of the action potential, outward current is driven across the potassium and chloride conductance because the depolarized membrane potential has moved farther away from the potassium and chloride equilibrium potentials. However, the inward sodium current overwhelms these outward ionic currents, and the vast majority of the outward current flows across the membrane capacitance, producing the depolarization.

The action potential peak never reaches the sodium equilibrium potential because the sodium inactivation gates close in response to depolarization (Figure 3.14c). Sodium inactivation gates have relatively slow kinetics—think of them as rusty. It takes more time for them to respond to depolarization than the sodium activation gates, so their closing does not prevent the rise of the action potential but does cut it short. Their closure reduces the conductance of the membrane to sodium, restoring the dominance of potassium, so the action potential peaks and then heads back toward the resting potential. The depolarization also turns on more potassium channel activation gates (Figure 3.4b), which also have slow kinetics. The opening of potassium activation gates of the half of the potassium channels that are closed at rest acts in concert with the closure of the sodium inactivation gates in bringing the action potential down toward the resting potential.

The repolarization of the membrane during the falling phase of the action potential resets the sodium and potassium channel gates to their resting configurations. Kinetics are the same for opening and closing—well-oiled gates are fast in both directions and rusty gates are slow in both directions. Repolarization quickly moves the fast-responding sodium activation gates toward their closed position, but full closure is at first obstructed by the closed sodium inactivation gates (Figure 3.14d), which do eventually open, getting out of the way (e), and allowing the sodium activation gates to snap shut (a), thus resetting the sodium channels to their resting potential configuration (see Kandel et al. 2013). Repolarization also resets the slow responding potassium activation gates to their resting configuration with about half of the channels open. Before resetting of the potassium activation gates is completed, there is a brief period when the relative permeability of the membrane to potassium is somewhat greater than at rest, causing the membrane potential to move closer to the potassium equilibrium potential, undershooting the resting potential before settling back to the resting level (Figure 3.13). It is important to remember that *fast* and *slow* are relative terms. The entire process is complete in just a few milliseconds and can occur hundreds of times per second in a single axon.

The energy expenditure during the action potential is reflected in the transfer of a small amount of sodium into the cell in exchange for an equal amount of potassium out of the cell. Because of the immense strength of the electric force, the amount transferred during a single action potential is far too small to produce a detectable change in the concentrations of these ions inside the cell, so the concentration gradients are unaffected. However, over the long term the sodium-potassium exchange arising from electrical activity is a significant energy expenditure, requiring about two-thirds of the metabolic activity of electrically active cells to run the sodium-potassium exchange pumps that maintain their internal potassium and sodium concentrations (Alberts et al. 2002).

The opening and closing of ion channels underlie all the activities of our nervous system. Because inward current entering the cell must always equal outward current leaving the cell, whenever conductance changes occur, whatever their cause, the results follow the same rules: inward ionic current produces outward capacitative current, which depolarizes the membrane, and outward ionic current produces inward capacitative current, which repolarizes the membrane. The changes in membrane potential produced by ionic currents across membranes are the fundamental processes underlying the function of our sensory systems, the nervous system, and muscle. Without membrane channels to produce these voltage changes we would not be walking around, appreciating a sunset, or considering life's meaning.

4

HODGKIN AND HUXLEY
BEFORE THE WAR

Jan Swammerdam (1637–1680) had the good fortune to pick the right animal for his investigations—the lowly frog. Then and now a relative nobody, he tested the properties attributed to animal spirits by Descartes (see Cobb 2002; Smith et al. 2012). His results should have pushed the animal spirits theory off the table into the trash heap of rejected scientific theory. An unlikely player in a major scientific drama, he was a reluctant medical student who obtained his M.D. in 1667 at the then ripe old age of 30, dumped science for religion just eight years later, and then died of malaria just five years after that at the age of 43. Somehow during his brief scientific career Swammerdam made important contributions. He established the frog nerve muscle as an experimental system that is used to this day. The advantage is that frog parts stay alive and functioning outside the body without special care a lot longer than mammal parts. He showed that touching a metal instrument to the cut end of a nerve extending from a muscle that had been removed from the frog caused muscle contraction even though there was no longer any connection to the brain. Without the connection, animal spirits flowing from the brain could not have caused the contraction. He enclosed a dissected frog muscle with the nerve intact in a glass vessel that he then sealed except for a thin tube projecting from the vessel near the top. A drop of water had been inserted in the tube before sealing the vessel. With this arrangement, just a tiny increase in the volume inside the vessel would be enough to cause the drop of water to move up the tube. Swammerdam mechanically stimulated the nerve with a wire that passed through a tightly sealed hole in the vessel. The muscle contracted, but the drop of water did not budge, demonstrating conclusively that muscle does not increase in volume even slightly during contraction (Turkel 2013). In a stroke of prophetic imagery, he characterized the signal that produced muscle contraction as rapidly propagating along a nerve, unaccompanied by the flow of matter, like a vibration conducted along a board. Swammerdam's results were not published until 1737, 57 years after his death. The human craving for the security of an explanation is strong, and theories are rarely abandoned unless another theory is waiting in the wings to be promoted by the scientific authority figures of the day. So belief that animal spirits caused muscle contraction lingered on, since there was no possibility of discovering the real mechanism without knowledge of electricity.

Swammerdam's results did not impress Herman Boerhaave (1668–1738), a renowned teacher of medicine at the University of Leiden in the Netherlands. He had strong religious and philosophical leanings and was a major proponent of animal spirits in the eighteenth century who still adhered to the classical mechanism that required animal spirits to flow from the nerves into muscles to make them contract by inflating them. From his base at the University of Leiden, Boerhaave exerted tremendous influence in science and medicine. Unlike Jan Swammerdam who performed decisive experiments but achieved little influence among the intelligentsia, Boerhaave appears to have achieved tremendous influence by quoting the work of others. In order to hold the view that animal spirits flow from nerve into muscle, Boerhaave dismissed

Swammerdam's work as only showing "that the Fabric of the Nerves in cold amphibious animals is different from that of the Nerves in Quadrupeds and hot Animals; so that no Argument of Force can be thence drawn to make any Conclusion with regard to the human Body" (Smith et al. 2012, 169). But at minimum Swammerdam showed that it is entirely possible for a muscle to contract after the nerve had been cut and without an increase in volume. Therefore, contraction without a nerve connection to the brain and without a volume increase must be considered a possibility for muscles in any animal, warm or cold, fuzzy or slimy. Because nerve and muscle fibers in frogs and people look similar, it is reasonable to go further and hypothesize that the mechanisms of nerve conduction and muscle contraction are similar in frogs and people and any other organisms with similar structures—not dismiss the possibility out of hand as Boerhaave did. One can only imagine how Boerhaave might have reacted to the discovery of the universal, electrical mechanism of the action potential in the squid axon in the twentieth century (see below and Chapter 5). You might ask how we really know that nerve conduction operates by the same mechanism in squids, frogs, and people. The answer is many ways, among which is that the same or very similar molecules are involved in the process across all species. Think of cars and trucks with gasoline engines containing cylinders, pistons, crankshafts, and many other parts that are essentially the same. Would it be reasonable to hypothesize that cars and trucks are powered by entirely different mechanisms?

Boerhaave's dismissal of the results of frog experiments as not applicable to warm-blooded quadrupeds is still relevant to science today. I recall a few years ago being asked by someone in a wheelchair if experiments with rat nerves have any relevance to helping people with spinal cord injuries, so the issue deserves some comment. Although more closely related than frogs, rats are not people either. However, the more fundamental the biological process, the more likely it is to have similar underlying mechanisms in all animals in which the process occurs. A prime example is the universal genetic code. Conduction of signals along nerves is a fundamental process common to all vertebrates and many complex invertebrate animals, so one might expect the underlying mechanisms to be very similar. As emphasized earlier, ideas about mechanisms are limited by the types of mechanisms available at the time; in the case of nerve conduction before the eighteenth century, all mechanisms were essentially plumbing. Swammerdam's observation that muscle contraction could be produced by contact of a metal instrument with the stump of nerve attached to a muscle that had been dissected out of a frog clearly showed that in the frog the connection of the nerve to the brain is not necessary to cause muscle contraction; that is, the mechanism in the frog *is not plumbing!* In the frog, something other than animal spirits flowing from the ventricles of the brain, through the nerve, and into the muscle caused the contraction—so why not in every animal with a nervous system? All the hypothetical mechanisms for nerve conduction up to and including Descartes involved assembling systems of plumbing, which included items such as hollow nerves, valves, and fibers, which were never observed, carrying the thinner-than-air animal spirits, which could not be seen or felt. Therefore, a credible example of nerve conduction in any animal that was not explainable by plumbing should have blown these houses of cards off the table. The only appropriate response for a scientist in that situation is to reexamine the existing theories without the constraints previously imposed by the requirements of plumbing mechanisms. But the powers that be are the powers that be for reasons that are not so easily swept away.

Unknown to Swammerdam, the metal instruments and metal wires he used to mechanically stimulate nerve actually produced what we now recognize as an electrical stimulus. Similarly, organs and tissues freshly isolated from the body can move in response to being touched with a metal instrument. This phenomenon became known as irritability. Albrecht von Haller (1705–1777), a Swiss-born physiologist working in Prussia, took the study of irritability to an extreme. Haller, along with an assistant, surveyed the irritability of tissues and organs in 190 living animals in a single year. We now know that muscle tissue is not confined to skeletal muscle and the heart. Muscle cells are of three types: skeletal muscle, cardiac muscle in the heart, and smooth muscle everywhere else. Smooth muscle is widely distributed, such as in arteries and veins, the walls of hollow organs, such as in the digestive tract, and the iris of the eye, to name just a few. All of the movement observed by Haller arose from the contraction of muscle. The movement of tissues removed from the body made it very clear that the tissues themselves produced the forces that caused the movements.

Now fast-forward to the twentieth century. Alan Hodgkin began his investigations into the mechanism of the action potential when he was an undergraduate student at Trinity College, Cambridge University, in 1934, but his first experiments look like Galvani (see Chapter 1) might have performed them back in the eighteenth century (Hodgkin 1992). A gastrocnemius muscle from a frog leg was strung up in a contraption to measure muscle twitches. The sciatic nerve, still attached, was stimulated with a pair of electrodes, and the muscle contraction recorded. What made the difference between Galvani's and Hodgkin's experiments was the accumulated knowledge of the intervening 150 years, but in 1934 the level of knowledge of the cell still had a long way to go. Electrons and ions were known to be carriers of electric current. The lipid composition of the cell membrane and the concentrations of ions inside and outside of the cell were known. The Nernst equation was known. Nobody could directly measure the resting potential, but it was known that cutting a nerve produced a negative voltage at the cut end, which suggested that the exposed inside of the axons is negatively charged relative to the outside. The prevailing hypothesis about the action potential mechanism, originated by Julius Bernstein in 1902, was that selective permeability to potassium produced a resting potential between −50 and −100 millivolts (mV), and the action potential was produced by a large increase in the permeability of the membrane to all ions that unleashed an inward ionic current, driving the membrane potential toward zero. Nothing was known of ion channels.

Hodgkin's undergraduate project would blossom into two papers, of which he was the sole author, comprising 50 pages in the *Journal of Physiology* (Hodgkin 1937a, 1937b). The papers were published when Alan Hodgkin was 23 years old. These were very different times. Nowadays, to make any splash at all, most biomedical scientists need to be pushing 40 or beyond and work as a member of a group, and there is no such thing as a single-authored biomedical research paper anymore.

PRIMITIVE EQUIPMENT

Hodgkin used the frog for the same reasons that Swammerdam and Galvani did—frogs were easy to obtain and keep, and their nerve and muscle tissue survived for many hours in a salt solution outside the body. In 1934 the state of the art was detection of action potentials using electrodes placed in contact with a nerve, but at the beginning of his undergraduate research project Hodgkin did not have access to an oscilloscope necessary for displaying and recording action potentials. He did inherit an apparatus for stimulating nerve with precisely timed shocks from Keith Lucas, a prominent physiologist at Cambridge 20 years before, and a smoked drum kymograph that could be used to record muscle contractions. The kymograph utilizes a mechanical linkage to connect the tendon at one end of the muscle to a stylus. The tendon on the other end is attached to a fixed point, and as the muscle contracts and relaxes the stylus traces the change in length by deflecting at a right angle to the direction of rotation of the drum. A motor rotates the drum at a constant rate, and the changes in contraction over time are recorded as a tracing in a coating of soot on the drum. The smoked drum kymograph is a higher-tech version of writing with a stick in the dirt.

This was enough to begin (see Figure 4.1). The motor axons contained within the sciatic nerve originate from cell bodies in the spinal cord and extend via the nerve to muscles in the lower limb. Each muscle is innervated by many axons. The frog gastrocnemius muscle in the lower leg (the one that can have painful cramps in humans) was the convenient choice of study. Each motor axon entering the gastrocnemius muscle divides into many small branches, each of which connects to a single muscle fiber, so each neuron innervates many muscle fibers. A strong electric shock delivered through stimulating electrodes placed in contact with the nerve causes action potentials to arise in the vicinity of the negative electrode in all the axons where the membrane is depolarized to threshold. The action potentials travel down the axons and invade the axon branches. The axon terminals release acetylcholine at synapses with the muscle fibers (see Chapter 6), causing action potentials to arise in the muscle fibers, resulting in contraction. The muscle twitches pull a lever that moves the stylus of the kymograph.

In his book *The Conduction of the Nervous Impulse,* Keith Lucas (1917, 13) nicely framed the major issues surrounding the characterization of the action potential in the early twentieth century: "The experiments are often easily made, even with a considerable degree of accuracy; it is in their interpretation that the real difficulty begins. And this difficulty arises again and again from the same cause, that nerves and muscles are

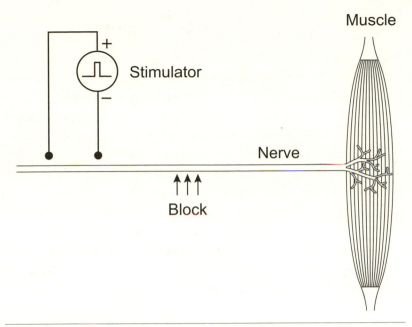

Figure 4.1. Depiction of Hodgkin's nerve-muscle experimental setup.

not units, but each composed of many fibres." This problem of interpretation comes up repeatedly in work with nerves containing multiple axons. Stronger shocks to the nerve produce stronger muscle contractions. One possibility was that a shock of higher voltage might produce a larger nerve impulse in an axon, which might in turn produce a stronger contraction of the muscle fibers the axon innervates. Another possibility was that nerve impulses are of fixed amplitude regardless of the strength of the stimulating shock, but that stronger shocks stimulate a greater number of axons to produce nerve impulses than weaker shocks, leading to the contraction of more muscle fibers and a greater strength of contraction. Lucas's book was published posthumously the year after he died in an airplane crash. He was flying as part of his job at the Royal Aircraft Factory where he worked on aircraft development at the beginning of World War I. Alan Hodgkin would very nearly follow in his footsteps while flying as part of his job developing airborne radar during World War II.

Working with multiple axons notwithstanding, Keith Lucas took a big step toward answering this question. As is often the case in any science, the trick is finding situations in nature where the phenomenon of interest can be more clearly observed, which makes the mechanism easier to reveal. Lucas just looked further within the frog at the cutaneous dorsi muscle, a small muscle in the back known to be innervated by fewer axons than innervate the gastrocnemius muscle—just 10 or less. He used an extracellular electrode placed on the surface of the nerve to stimulate the axons. Axons in a nerve have different firing thresholds when stimulated by extracellular electrodes because axons farther away from the negative electrode receive a smaller share of the stimulating current than axons that are closer, and therefore a stronger shock is needed to bring them to threshold. As Lucas increased the voltage of the shock applied to the nerve, the strength of muscle contraction increased to its maximum in several steps, always fewer than 10. The logical explanation was that as the stimulating voltage increased, there was no change in the strength of contraction until the voltage reached the firing threshold of the next axon. Then the strength increased in a jump, adding the contributions of the muscle fibers innervated by the additional axon to the power of the contraction. The steps of increased contraction could be smaller than the number of axons because sometimes more than one axon would have about the same threshold. The fact that the contraction did not increase at all between the steps suggests that the size of the nerve impulse is independent of the stimulus strength.

Much evidence for what we now refer to as the all-or-none character of the action potential came from experiments in which frog nerve was arranged to pass through a chamber containing an anesthetic used to block the transmission of the nerve impulse. This was, in fact, the principal means of investigating the nature of the nerve impulse in the early twentieth century (Figure 4.1). It was surprisingly informative. Several investigators published experiments using this method to address the question of the amplitude of the action potential. Among them, the experiments by Adrian published in 1914 (Lucas 1917) were especially decisive. Edgar Adrian, perhaps better known to us as Lord Adrian, had become Master of Trinity College by the time Hodgkin was a student there. When nerve impulses traveling along a frog nerve reach a chamber that contained anesthetic, the passage of the nerve impulse is blocked if the section of nerve in the chamber is long enough, about 9 millimeters (mm). If the anesthetized section of nerve is half that length, 4.5 mm, the impulse passes through and the muscles contract the same as when no anesthetic is used. One possibility was that passage through the longer segment diminished the amplitude of the nerve impulse to the point where it failed, while nerve impulses emerging from the shorter segment were still able to carry on at a reduced amplitude, but still strong enough to trigger full contraction of the muscle. Adrian hypothesized that if the nerve impulses emerging from a 4.5-mm anesthetic-blocked segment of nerve were diminished in amplitude, another 4.5-mm anesthetic-blocked segment encountered further along the axon should finish them off, and conduction should grind to a halt. Simply stated, two 4.5-mm segments of anesthetic-blocked axon separated by a stretch of conducting axon should be just as effective in blocking transmission as a 9-mm stretch of anesthetic-blocked axon. Adrian's experiments showed that it is not. The impulse emerging from the first blocked region also blasts right through the second. In his description of Adrian's experiments, Keith Lucas (1917, 23–24) aptly compares the conduction of the nerve impulse to a train of gunpowder (I prefer a dynamite fuse), "where the liberation of energy by the chemical change of firing at one point raises the temperature sufficiently to cause the same change at the next point. Suppose that the gunpowder is damp in part of the train; in this part the heat liberated will be partly used in evaporating water, and the temperature rise will be less, so that the progress of the chemical change may even be interrupted; but if the firing does just succeed in passing the damp part, the progress of the change in the dry part beyond will be just the same as though the whole train had been dry."

From our modern biological perspective it is hard to imagine it any other way. Biological tissues simply do not possess the level of conductivity of metal wires required for a localized voltage source to project a strong-enough signal at a distance without amplification. In order to travel the distances in the body that are required, self-propelled action potentials would seem to be necessary. Because the axon provides the energy that propels the action potential all along its path, an action potential arriving at a branch point in the axon generally results in full-blown action potentials in both branches (branch points in axons are normally bifurcations). Because of this, in the brain and spinal cord a single action potential generated at the axon where it emerges from the cell body can multiply at thousands of branch points, spawning thousands of action potentials penetrating into the thousands of axon terminals.

BEFORE THE SQUID

All of these experiments were known to Alan Hodgkin, and he even had the very same stimulating apparatus that Keith Lucas had used. It had been known since the time of Galvani that electrical stimulation could produce nerve impulses in frog nerve, and it was well known in Hodgkin's time that the action potential is accompanied by an electrical disturbance in the nerve. This observation would have made Galvani jump for joy, but many still considered the electrical nature of the nerve impulse an open question. I suppose that to some it just did not seem right that nerves should carry electrical signals. It seemed more likely that a tiny flow of electricity was an accompaniment of the nerve impulse that did not participate in its mechanism of travel. In science there is a derogatory term, *epiphenomena,* used to refer to phenomena (usually investigated by one's competitors!) that do not participate in mechanisms under investigation and seem just to be there to confuse scientists. Hodgkin, undeterred, set out to address the issue of the electrical nature of the action potential in his undergraduate research project at Trinity College.

If Bernstein's 1902 theory was correct, the axon membrane becomes permeable to all ions during the action potential, and so the action potential should be accompanied by a large increase in the electrical conductance of the membrane as it travels along the axon. Hodgkin designed an experiment to detect the predicted increase in conductance. Membrane conductance would become easy to measure directly once Hodgkin and Huxley worked out how to record with intracellular electrodes inserted inside a squid axon years later, but detecting an increase in conductance in the axon by measuring the strength of muscle con-

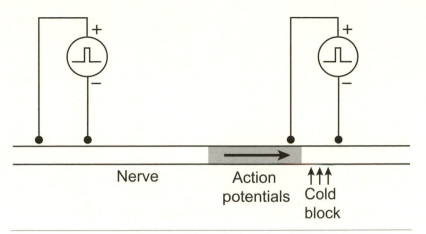

Figure 4.2. Stimulation with the positive electrode above the cold block.

traction required some mental gymnastics. Hodgkin's approach, which was modeled after the work of Lucas and others, is interesting because it illustrates how nerve impulses were studied during most of the first half of the twentieth century. Hodgkin set up the frog sciatic nerve in an apparatus that allowed a short segment of the nerve to be locally chilled to the point of blocking the conduction of action potentials (Figure 4.2). During the cold block, the length of nerve above and below the cold block retained the ability to conduct action potentials. For the experiment to be meaningful the cold block had to be reversible; conduction through the blocked region had to be restored by warming to show that the cold-blocked segment of the nerve was not just dead.

Firing nerve impulses at the cold block was accomplished with two stimulating electrodes placed in contact with the nerve a distance upstream. The shock used was strong enough to generate action potentials in all the axons in the nerve. Hodgkin set up another pair of stimulating electrodes across the cold-blocked region of the nerve, with the positive electrode just above the cold block and the negative electrode just below. Stimulating current flowing between the positive and negative electrode takes several paths. Most of the current takes the path of least resistance from the positive electrode through the extracellular fluid to the negative electrode and has no impact on the nerve. A minority of current travels the path of greater resistance, across the membrane into the axons, through the cytoplasm inside the axons (known as the axoplasm),

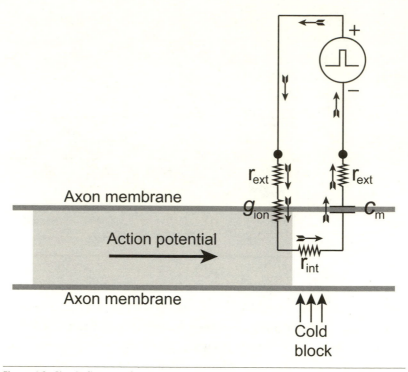

Figure 4.3. Circuit diagram of a Bernstein action potential arriving at a cold block spanned by a stimulating electrode.

and out across the membrane to the negative electrode. When crossing the membrane, current goes through the ion channels and the membrane capacitance. Figure 4.3 shows the path of stimulating current that is relevant to Bernstein's theory. Flowing counterclockwise in the circuit diagram, the current crosses the resistance of the external fluid (r_{ext}) from the electrode to the axon, enters the axon through the ion channel conductance (g_{ion}), travels forward along the axon through the internal resistance (r_{int}), exits through the membrane capacitance (c_m) causing a depolarization, and finally travels through the resistance of the external fluid to the negative electrode. The current entering the axon actually flows in both directions along the axon, but only the current flowing along the axon into the cold-blocked region is relevant to the experiment.

It takes a certain amount of outward current through the membrane capacitance to bring the membrane to threshold and fire an action potential. When stimulation is applied through the electrodes across the cold block, an increase in conductance anywhere along the path of the stimulating current will cause the current flowing through the entire path to increase. Hodgkin attempted to use this arrangement to detect if the arrival of action potentials just above the cold block causes an increase in the ionic conductance in the membranes of the axons in the nerve as predicted by the Bernstein theory. To make his experiment as sensitive as possible in detecting an increase in conductance just above the cold block, Hodgkin arranged the length of the nerve exposed to the cold block to be just long enough to block the action potentials in all the axons in the sciatic nerve from crossing it. Then he adjusted the strength of the shock delivered across the cold block to a voltage just below the threshold; just a little bit more current was needed to generate action potentials below the cold block in the axons with the lowest threshold and produce a detectable twitch of the gastrocnemius muscle. Hodgkin fired off action potentials from the upstream electrodes. Then at exactly the time the action potentials arrived just above the block, he stimulated the nerve with the just-subthreshold shock across the cold block. The once-subthreshold stimulation now produced a contraction of the gastrocnemius muscle. Voilà, the arrival of action potentials must have increased the membrane conductance (g_{ion} in Figure 4.3) just above the block, which amplified the stimulating current crossing out of the axon through the membrane capacitance (c_m) just below the block, depolarizing the membranes of some of the axons just below the block to threshold, producing action potentials that triggered contraction of the gastrocnemius muscle.

But, not so fast! In Hodgkin's own words: "Then, after five or six weeks I had a horrid surprise. I switched the positive electrode from just above the block to a position [along the nerve] beyond it   and found that the effect persisted" (Hodgkin 1992, 64–65). With the positive electrode also below the block (Figure 4.4), the effectiveness of stimulating the axons was still increased by the arrival of action potentials at the block. The reduced threshold for the production of action potentials by stimulation just below the block had nothing to do with the path of the stimulating current through the membrane just above the block. Hodgkin's horror was based upon having made a mistaken interpretation based on the initial lack of an important control experiment. In fact, the position of the positive electrode does not much matter when stimulating a nerve, as long as enough outward current

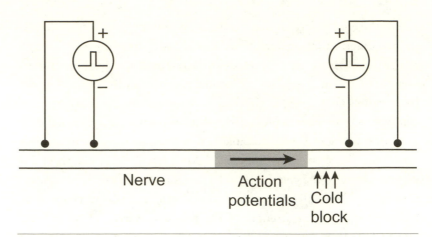

Figure 4.4. Stimulation with the positive electrode below the cold block.

crosses the axon membrane in the vicinity of the negative electrode to depolarize the membrane to threshold. I am reminded of one of my favorite movie quotations. This one is from *The Ghost and the Darkness* where the professional big-game hunter Remington (Michael Douglas) is consoling the engineer John (Val Kilmer) who was almost killed by a man-eating lion because the borrowed rifle that he had not tested had misfired: "They['ve] got an expression in prizefighting: 'Everybody's got a plan until they['ve] been hit.' Well, my friend, you['ve] just been hit. The getting up is up to you." (Good advice, I think.) Many times the results of experiments send experimenters in a different direction from their original intention. Much of success depends upon how they react.

In fact, the axons themselves were doing just what Hodgkin thought he was doing with the stimulating electrodes straddling the block. Because of Hodgkin's work with Alan Huxley years later we now know that the inward current associated with the action potential is produced by the opening of sodium channels in the axon membrane, not by an increase in the permeability to all ions as proposed by Bernstein. Sodium driven into the axon by its concentration gradient powers the active rise of the action potential. A portion of the inward sodium current spreads forward inside the axon, exiting across the membrane and looping back to complete what has become known as a local circuit. This current loop depolarizes the membrane ahead. In Hodgkin's experiment, it is the active rising phase of the action potential that is stopped at the cold

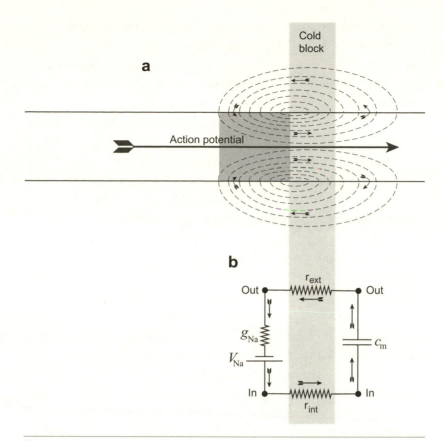

Figure 4.5. Current loops extending ahead of the region of active rise of an action potential that has reached the cold block: (a) forward current loops; (b) circuit diagram of the forward current loops.

block, because cold prevents the sodium channels from opening fast enough to initiate the positive feedback cycle that generates the depolarization. But cold does not prevent the spread of current through the cold-blocked section of the axon, so the passive rise of the action potential penetrates the cold block depolarizing the region ahead, thus lowering the threshold for stimulating action potentials at a negative electrode located just below the block. It doesn't matter at all where the positive electrode is located.

Figure 4.5a depicts the current loops extending forward from the active rising phase of an action potential that has reached the cold-blocked region of a single axon. Some of the current entering the axon in the region of the active rising phase (shaded) stays within the axon and projects forward, passing through the cold-blocked region, and exiting the axon across the membrane capacitance just below the cold block. This current depolarizes the membrane beyond the block, but not quite to threshold. In Hodgkin's experiment, the additional depolarizing current provided by the negative electrode located below the block produces the additional depolarization that brings some of the axons to threshold.

Thus, the action potential travels along by acting as its own stimulator. You can see from the circuit diagram of the local circuit in Figure 4.5b that the sodium battery in one location, in this case in the region of the active rising phase of the action potential, acts like an electrical stimulator to produce an outward current across the membrane at a location ahead, thus depolarizing it.

Take a really good long look at the local circuit in Figure 4.5b. This circuit not only explains conduction along the axon, but with minor variations it also functions in communication between neurons in circuits in the brain, learning and memory, and just about every other manifestation of animal electricity. All these functions have at their core ion channels that generate electric current that flows through the internal resistance of the cell to change the voltage across the membrane capacitance at a distance away. We will see in Chapter 6 that this circuit is the reason neurons look the way they do. Hodgkin stumbled across something great in his experiment designed to look at something else.

Hodgkin needed a means to directly record the electrical changes in the nerve associated with the action potential in order to directly test the local circuit theory. The problem with recording electrical events is their speed. Muscle contraction is much slower, so a smoked drum kymograph is adequate. Faster electrical signals can be recorded with this kind of apparatus by moving the stylus with an electromagnet driven by the amplified electrical signal, but moving the mass of the linkage and stylus is still too slow to record an action potential, which would be long gone before the inertia of the system was overcome and it started moving. Hodgkin had access to a Matthews oscilloscope, an electrically driven mechanical device that minimized mechanical inertia by using the amplified signal to drive small movements of a mirror that deflected a beam of light from an arc lamp projected on a rotating cylindrical paper screen—a maximally refined version of the smoked drum. Hodgkin could detect a deflection when the action potential passed by the recording electrodes, but the system was still too slow to get an accurate tracing of the shape of the action potential. Ultimately he got his hands on a primitive cathode-ray tube and constructed an oscilloscope that could display fast signals such as an action potential. It was perhaps not too difficult for him to get expert advice since J. J. Thomson, who led the group that discovered the electron in 1897, was Master of Trinity College at the time.

Hodgkin, along with everyone else working on action potentials at the time, was limited to extracellular recording of action potentials that occurred simultaneously in many axons within the nerve. While action potentials in individual axons are of constant amplitude, simultaneous action potentials traveling in the nerve add together to produce a "compound action potential" in the nerve made up of the summed amplitudes of the action potentials in many axons. A submaximal shock to the sciatic nerve produces a compound action potential of lower amplitude than a maximal shock because fewer axons are firing, and so the amplitude of the compound action potential can be used to determine changes in the number of axons firing, just like measuring muscle contraction. By the middle of 1936, Hodgkin had obtained electrical recordings of action potentials in the frog nerve subjected to cold block that supported the idea that electric current spreading ahead of the active rising phase of the action potential is responsible for propagating the action potential forward. The arrival of the action potentials caused a change in electrical potential to be projected across a 2–3-mm cold block that was accompanied by an 80–90% reduction in firing threshold in the region just beyond the cold block.

At some point Adrian remarked to Hodgkin that he should study crab nerve. Hodgkin took him up on the suggestion since crab motor nerves lack the outer sheath, called the perineurium, which is present in all mammalian nerves and can interfere with experimental treatments and measurements. Axon membranes alone have little structural strength, but axons have structural proteins called neurofilaments contained within them, and they are surrounded by thin sheaths of connective tissue, both of which provide some structural strength. Using only the naked eye, Hodgkin was able to tease apart a crab nerve into fine strands. He placed them one at a time in his apparatus, electrically stimulated them with increasingly strong shocks, and observed the resulting action potential displayed on his oscilloscope. Most strands contained several axons and displayed compound action potentials whose amplitude increased in steps as the shock strength was increased, bringing additional fibers to threshold, but in one case the stimulation resulted in an "enormous all-or-none action potential" (Hodgkin 1992, 77). Hodgkin knew at once that by chance he must have pulled out one of the six or so very large axons in the crab nerve with a diameter of about 30 micrometers (μm). Hodgkin knew that he now had a way to get around the problem of recording from multiple axons in nerves that had complicated the interpretation of everyone's experiments in the past. He borrowed a dissecting microscope so he could always find single crab axons, and he pretty much bid what must have been a fond farewell to the frog and multiple axons forever.

Before Hodgkin could get much done with crab nerves, he headed to New York on a traveling fellowship granted by the Rockefeller Institute. During his travels he would meet many of the greats of American neurophysiology. Among them were Joseph Erlanger and Herbert Gasser of Washington University in St. Louis, who would share the Nobel Prize in Physiology or Medicine in 1944 for their discoveries made working with single nerve fibers. Erlanger and Gasser had read Hodgkin's undergraduate thesis, and Erlanger wrote to him in 1937: "In general I find it hard to believe that nerve impulses are propagated by currents eddying outside of the conducting structure. Teleologically such a mechanism seems queer. The fact that we have found it impossible to demonstrate in an intact nerve any alteration in the excitability of inactive fibers lying parallel to active fibers lends support to our skepticism" (Hodgkin 1992, 74).

The idea that the electric currents that flow outside of the axon during an action potential (Figure 4.5a) might stimulate the axon next to it to produce an action potential was to become a significant issue. We now know that this does not happen. The current would have to cross from the generating axon into and out of the neighboring axon, and then back to the generating axon. The additional resistance provided by two passages across the membrane of the neighboring axon makes this path much higher in resistance than the current path through the extracellular fluid, so only a tiny portion of the current takes this path—much too little to depolarize the neighboring axon very much, and nowhere near enough to bring it to threshold. Thus, axons do not need to be insulated very much from one another to prevent cross talk. Likewise, the vast majority of the current passed between two extracellular stimulating electrodes travels just through the extracellular fluid, not into and out of the axon being stimulated. Stimulation with extracellular electrodes can produce action potentials because a relatively strong shock is used to drive a relatively large current, and the small fraction of the current that does cross the membrane is enough to depolarize the axon to threshold.

Hodgkin visited St. Louis during his traveling fellowship. Upon meeting Hodgkin, Joseph Erlanger issued a friendly challenge: he would take the local circuit theory seriously if Hodgkin could show that altering the electrical resistance outside of the nerve changed the conduction velocity. This would be a good experiment because the lower the external resistance the more current travels in the forward current loop, and that current extends farther ahead along the axon, depolarizing the membrane farther ahead, thus increasing the conduction velocity. During normal conduction of an action potential, the major resistance to the forward current comes from the resistance of the axoplasm (r_{int} in Figure 4.5b) confined within the axon, which is like a very fine, high-resistance wire. The resistance of the extracellular fluid surrounding the axon (r_{ext}) is always much lower than the internal resistance because of the much larger cross-sectional area of the external solution through which the current flows during its return.

You cannot much change the internal resistance, so in accepting the challenge Hodgkin devised ways to both increase and decrease the resistance of the external current paths surrounding an axon. The species of live crabs available from markets in New York City had nerves that were too delicate, and it was hard to isolate functional axons from them, so he had some live specimens of *Carcinus maenas,* the species whose axons had worked so well in Plymouth, shipped to him on the *Queen Mary.* Hodgkin increased the external resistance by floating a thick layer of oil on top of the saltwater bathing the isolated axon. Stimulating electrodes were placed at one end of the axon and recording electrodes at the other, and the conduction time was measured on the oscilloscope trace between the stimulus artifact and the rise of the action potential. The axon was lying loosely between the electrodes over a movable hook so that the axon could be gently raised into the oil without damaging it. When raised into the oil, the axon retained a thin coating of salt solution, which was enough extracellular fluid to permit the conduction of action potentials, but was of much higher resistance to the flow of current along the axon than the large volume of solution from which the axon had been raised. True to prediction, the conduction velocity decreased, about 30% as it turned out, when the axon was raised into the oil. In Alan Hodgkin's words: "This was one of the few occasions on which everything went according to plan, and this time no hidden snag emerged. I showed Harry Grundfest the records next day and remember that he shook me by the hand, like a character in a novel by C. P. Snow" (Hodgkin 1992, 113). Grundfest, a famous neurophysiologist who made many contributions, collaborated with Gasser on some of the early work on nerve transmission.

THE SQUID GIANT AXON

The field of animal electricity was about to be transformed by the discovery that ordinary squids have giant axons with diameters of around half a millimeter. As stated retrospectively by one of the major scientists who studied squid axons, Richard Keynes (2005, 179), "It was, however, the introduction of the giant nerve fibres by J. Z. Young that enabled the biophysics and biochemistry of excitable membranes to be properly studied in depth, which was said by Alan Hodgkin in 1973 to have done more for axonology than any other single advance in technique during the previous 40 years." The animal that stands out as the greatest gift to the fundamental understanding of how the nervous system works is not even a vertebrate.

As you read further you may wonder why I say this since many more experiments have been performed using frogs, and the mechanism of synaptic transmission was first revealed in experiments with the frog nerve-muscle system. Frogs were used because of their many conveniences, especially the ease with which frog tissue can be kept alive outside the body. The same experimental analyses were eventually performed using mammalian nerve-muscle systems without the conveniences offered by working with frogs. Without the frog, progress would have been a bit slower. The squid is unique because its giant axons offer the opportunity for direct tests of the mechanisms of generation and conduction of action potentials. To this day no one has discovered an axon in any other animal big enough to permit an electrode to be slipped in through the cut end and snaked down its entire length. Without the squid, the knowledge of the mechanism of conduction would have been obtained from somewhere, likely pieced together from multiple, less direct sources, but I think it might have been a very long time coming.

In 1936 a British zoologist, John Zachary (J. Z.) Young, published his discovery that certain squids have giant axons (Keynes 2005). Thus, the development of our knowledge of how the nervous system makes and conducts electrical signals in all animals with nervous systems including humans was largely enabled by a discovery made by a zoologist. This was two centuries after the discovery of animal electricity. In the squid giant axon the mechanism of conduction of the nerve impulse could be revealed as in no other animal. The torpedo ray and electric eel showed that animal electricity really exists. The squid giant axon allowed scientists to figure out what animal electricity really is.

Giant axons are important to the squid. They connect the squid's brain to the muscles of the squid's spindle-shaped body, its mantle. The mantle forms a tube composed of the thick muscular wall that tends to get turned into calamari rings after humans catch the squid. The mantle tube is closed at the tail end and open to the seawater at the head end. There are three openings, a narrow funnel and lateral openings, one on each side of the funnel. Expansion of the mantle increases its volume, drawing water in through the lateral openings where it can provide oxygen to the gills, and contraction of the mantle reduces its volume, exhaling the water. In addition to breathing, many species of squid use their mantle for jet propulsion. The squid squeezes the seawater-filled mantle by contracting the muscle fibers that encircle the mantle cavity and simultaneously closes the lateral openings of the mantle, forcing the seawater rapidly out the small opening of the funnel in a powerful jet, propelling the squid in the opposite direction. The squid can aim the funnel forward or backward in the opposite direction of the way it wants to go. This is important to the squid because it is how the squid escapes predatory fish that want to eat it.

Once it has decided to escape, the squid must send the message rapidly from its brain to the muscle fibers of the mantle. To provide maximum propulsive force, all the mantle muscles must contract simultaneously so that the mantle is squeezed all at once to squirt seawater out of the funnel at high velocity. The conduction velocity of the action potential in squid axon is about 20 meters per second, which is fast but nowhere near as fast as the near light speed at which electricity is conducted along a wire. At this conduction velocity the need for simultaneous contraction of the mantle muscles is an issue, because the mantle muscles that ring the mantle near the tail of the squid are farther from the brain than the mantle muscles near the head of the squid. So when the squid decides to escape, sending action potentials down the axons, the action potentials traveling along nerve branches connected to muscles near the tail have farther to go than the action potentials traveling along nerve branches connected to muscles near the head. If action potentials traveled at the same velocity along all the nerve branches, the muscles near the head would contract before the muscles near the tail, and while some of the seawater in the mantle would be propelled out of the funnel, some would move inside the mantle cavity toward the tail where it would have to wait until the muscles near the tail contracted to move the water back toward the head and out through the funnel. So the force of propulsion would not develop as rapidly, and squids with this kind of system would wind up in fish stomachs more frequently than squid that could squeeze their mantles with maximum force achieved by simultaneous contraction of all of the muscles. There would be no issue if axons conducted action potentials at the speed of light; the mantle length would not affect the arrival times of the action potential. But mantle length is an issue because of the much slower velocity of action potentials. If wires conducted electrons at the speed of an action potential, landline telephones could not exist.

The way that squids have solved the problem of simultaneous contraction of the mantle is based on the physics of the action potential. It turns out that because of the laws of physics, nerve action potentials travel faster along larger-diameter axons than along smaller-diameter axons. You will learn why in Chapter 5. Vertebrates have achieved high conduction velocities in small-diameter axons with a layer of electrically insulating fat, the myelin sheath. How this works is complicated and also discussed in Chapter 5. Axons in invertebrates, such as the squid, do not have myelin sheaths. Instead, they achieve high conduction velocities by having axons with extremely large diameters, approaching 0.5 mm, which is 25 times larger than the 20-μm diameter of the largest mammalian motor axons. Since the cell bodies of neurons are very much smaller than 0.5 mm in diameter, and since the axons depend upon the cell body of the neuron to synthesize many of their proteins and other components, it is impossible for a single cell body to produce and maintain a squid giant axon. Instead, small fibers from many cell bodies in the stellate ganglion of the squid are fused together during the embryological development of the squid to form a large axon. The cell membranes of the contributing neurons form a continuous surface membrane so that the giant axon and all of its connected cell bodies function as a single cell. Clearly, the existence of such a complex system indicates that the large diameter of the fibers serves an important function—in this case, rapid escape from predators.

Axons from the squid's brain connect with the stellate ganglion located at the head end of the mantle, and the motor fibers radiate out from the stellate ganglion (hence its name) to innervate muscle fibers from the head end to the tail end. In the squid's mantle, the axons that connect to the muscle fibers at the tail end are larger in diameter than the axons that connect to the muscle fibers at the head end, and there is a progression of increasing axon diameter along the mantle from head to tail. Because of this, the action potentials that have the farthest to go to reach the muscles in the tail end of the mantle have faster conduction velocities than the action potentials that have shorter distances to reach muscles nearer the head end of the mantle. Because of the gradient of increasing diameter of the innervating axons along the mantle, and the resulting increase in conduction velocity with increasing axon length, the nerve impulses departing simultaneously from the stellate ganglia reach the nerve endings simultaneously all along the mantle, causing the mantle muscles to contract simultaneously, squeezing the mantle all at once, producing a high-velocity jet of water out the funnel, and the squid escapes the fish. Because it is best for this system to work as fast as possible after the squid makes the decision to escape, all of the axons in the mantle are relatively large, and the axon that connects to the tail end is the biggest of all, appropriately called the giant axon.

Two major centers of research developed in locations where squid were accessible—one in Plymouth, England, and the other in Woods Hole, Massachusetts. Working with squid had its disadvantages since they do not survive very long after capture, so squids netted by trawlers early in the day were used immediately in experiments that sometimes lasted into the small hours of the next morning. However, summers by the sea (with a constant supply of free calamari?) likely offered some consolation.

Hodgkin spent the summer of 1938 at the Marine Biological Laboratory at Woods Hole. Howard Curtis and Kenneth Cole (better known as Kacy Cole), who were the major investigators of the action potential mechanism in the United States, were working at Woods Hole, and they let Hodgkin use their amplifier to experiment with squid axons. Raising the squid giant axon into oil slowed the conduction velocity, just as it had with crab axons. Hodgkin was also able to greatly reduce the external resistance around the axon and determine its effect on conduction velocity. The length of squid axon between the stimulating and recording electrodes was laid across a series of platinum strips oriented crosswise to the axon and not touching each other. The axon was not immersed in salt solution, but the salt solution adhering to its surface was kept from evaporating by keeping the chamber housing the axon moist, which was enough to maintain the ability of the axon to generate action potentials during the experiment. The opposite ends of the platinum strips extended downward over a container of mercury, which could be raised, immersing the ends of the platinum strips and thus electrically connecting them together without disturbing the axon.

With the ends of the platinum strips immersed in mercury, the path of external current along the axon was entirely through metal, a much better conductor than the thin coating of salt solution on the axon. Hodgkin (1992, 115) describes the result: "When the strips were connected together there was the expected increase in conduction velocity. Such an experiment affords strong evidence because the only agent known which could travel through a metallic short-circuit in the time available is an electric current." As strong as this evidence was, it would still be some years before the local circuit theory was accepted.

Cole and Curtis (1939) made some key basic measurements working with the squid giant axon that summer. By placing electrodes transversely across a squid axon, which was lying in a narrow, seawater-filled channel, they were able to get a relative measure of the membrane conductance because a large component of the current that was conducted between the electrodes was forced to pass through the axon membrane twice. Cole and Curtis established an alternating current between the electrodes, which allowed them to observe changes in conductance, then fired an action potential using stimulating electrodes placed at a distance from the transverse electrodes. When the action potential passed between the transverse electrodes the membrane conductance increased greatly, by about 40 millisiemens per square centimeter of membrane at the peak of the action potential. They also estimated the membrane capacitance at about one microfarad per square centimeter. Capacitance depends upon the thickness of the membrane, which is fixed and remains constant during the action potential. It took knowledge of electricity far beyond Ohm's law and much mathematical gymnastics to produce these figures, but such measurements would soon be simplified by the advent of the intracellular electrode, which allows direct recording of the membrane potential and direct passage of current across the membrane. That summer Curtis and Hodgkin working together attempted to insert electrodes into the cut end of the squid giant axon without success, but they believed that it would ultimately be possible after a little more troubleshooting. Both would succeed on different sides of the Atlantic the very next year.

In October 1938 Hodgkin was back in Cambridge putting together the equipment to launch his career as an independent investigator. He muses that as neurophysiology became a recognized profession the objective in outfitting a laboratory with electronic equipment involved making everything look as complex as possible as a way of "cowing your scientific opponents or dissuading your rivals from following in your footsteps" (Hodgkin 1992, 124).

Hodgkin was soon required to embark on his teaching career—a prospect that many academic researchers dread. His first teaching assignment was a first-year laboratory class. One of the students was Andrew Huxley, whom he would eventually recruit as a collaborator. Hodgkin confesses to making the usual mistakes in beginning his teaching career of preparing too much material and focusing on the facial expressions in the audience—something most of us who have lectured can identify with.

Hodgkin's experiments in Cambridge in 1938–1939 immediately got off to an interesting start. He began working with crab axons again and was joined by Andrew Huxley, who was by then a more advanced physiology student. The standard method for measuring the membrane potential by extracellular recording was to place an external electrode at the membrane along the axon to detect the external voltage and estimate the internal voltage with an electrode placed at the cut end of the axon where the membrane was interrupted. This arrangement would always yield a voltage less than the real membrane potential because the "internal" electrode was not really inside the axon, so there was a short circuit between the two electrodes through the bathing solution that did not pass across the cell membrane. Nonetheless, Hodgkin and Huxley were able to measure the resting membrane potential and the membrane potential at the height of the action potential accurately enough. Bernstein's 1902 theory was still the prevailing theory of how voltages arise across the axon membrane: The resting membrane was believed to be selectively permeable to potassium, and so the potassium concentration gradient produced a resting potential with the inside negative. During the action potential, the selective permeability of the membrane to potassium was swamped by an increase in membrane permeability to all ions, that is, a breakdown in the selective ionic permeability of the membrane. This would drive the membrane potential to approach zero at the crest of the action potential. But Hodgkin and Huxley found that action potentials were larger than the resting membrane potential—a lot larger. For example, an axon with a -37 mV resting potential produced an action potential 73 mV in amplitude. This means that the membrane potential passes through zero, peaking at $+36$ mV. Such a reversal in polarity of the membrane potential is not possible according to Bernstein's theory. It was becoming imperative that a method be devised to make direct measurements of the membrane potential with an electrode inserted into the axon.

In July 1939 Hodgkin packed up his humongous load of competitor-terrifying gear into a trailer he had purchased for that purpose and towed it to the Plymouth Laboratory where squid were available fresh from the ocean. He and Andrew Huxley set up for the summer's research. Huxley was a commensurate tinkerer who liked to set up cute, unlikely experiments. He got it in his head to measure the viscosity of axoplasm by hanging an axon vertically and inserting a droplet of mercury into the cut end at the top and watching its rate of fall within the axon using a horizontally mounted microscope. It turned out that the droplet did not fall at all, and Huxley humorously concluded that axoplasm must be a solid. We now know that axoplasm is, of course, liquid, but it also has within it protein structures invisible through the light microscope unless specifically stained. Among these are neurofilaments, which give structural strength to the axon, and microtubules, which are part of an axonal transport system that carries molecules in both directions along the axon. Without the axonal transport system axons could never grow and form circuits. Knowledge of this would come many years later.

Apparently looking at the axons hanging vertically during this failed experiment produced an "aha moment." In the words of Alan Hodgkin (1992, 133): "Huxley said he thought it would be fairly easy to stick a capillary down the axon and record potential differences across the surface membrane. This worked at once, but we found the experiment often failed because the capillary scraped against the surface membrane; Huxley rectified this by introducing two mirrors which allowed us to steer the capillary down the middle of the axon."

The capillary electrode was just a very fine glass tube filled with a conducting salt solution, open at the end inside the squid axon, and with a wire inserted into the other end. This creates a path for current directly into the axoplasm so that the membrane potential can be directly measured. Hodgkin and Huxley recorded resting potentials of about −50 mV and action potentials with amplitudes approaching 100 mV, blowing Bernstein's 1902 theory out of the water.

It would be hard to imagine a more exciting result, especially for two young people, but the next step would be a long time coming because three weeks later Hitler invaded Poland. The Plymouth Laboratory collecting vessel was immediately commandeered as a minesweeper. Hodgkin and Huxley would both wind up in the armed forces working on radar, although in different venues. They would not return to research on squid axons until 1947. Hodgkin and Huxley did manage to publish a brief note in the journal *Nature* on the overshooting action potential in squid axon in 1939. In that paper they did not even speculate that an increase in sodium permeability might be responsible for the overshoot, although sodium was the obvious candidate. Writing his 1992 autobiography, Hodgkin could not recall why. While Hodgkin and

Huxley were occupied with the war effort, Curtis and Cole at Woods Hole published a more extensive study of the overshooting action potential in squid axon in 1942, which contained the confusing result that squid axons bathed in isotonic sucrose solution lacking all ions, instead of seawater, still conducted action potentials. This was reported as a single line in their paper with no data and no detail (Curtis and Cole 1942).

Immersion of an axon in sucrose dissolved in water is different from immersion in oil. Immersing an axon in oil does not disperse all the water and ions associated with the polar heads of the membrane phospholipids, so the external conducting solution remains as a thin layer. Sucrose was added to the water in the concentration needed to produce an isotonic solution, which provided the same number of dissolved particles per liter as present inside the axon and thus exerted no osmotic pressure that could have caused damage. Most of the ions associated with the outside of the axon would likely have diffused away into the sucrose solution, becoming too dilute to participate in the action potential mechanism. So, in principle, Curtis and Cole's experiment looked like a good test of the sodium theory and local circuit theory, both of which seemed to have been disproved by the persistence of the action potential when the axon was bathed in the sucrose solution.

While Curtis and Cole's result was correct, it turned out to be irrelevant. Curtis and Cole did not realize that as in the case of vertebrates a thin layer of tissue, the perineurium, covered the squid axon, which retained a coating of ionic solution in squid axons that was not immediately accessible to the sucrose solution. Just as in the case of axons immersed in oil, this layer contained enough sodium to maintain the action potential mechanism during their experiment. But at the time this misinterpreted result drew the focus away from sodium as the possible carrier of the inward current during the action potential. There were still those who believed that the electrical events associated with the action potential were consequences of the action potential but not part of the actual mechanism—the dreaded epiphenomenon.

5

THE MYSTERY OF
NERVE CONDUCTION
EXPLAINED

Even though work on the squid axon at Woods Hole continued uninterrupted during World War II, the discovery of the mechanism of the action potential would await the return of Hodgkin and Huxley to their laboratory. Once back, they would solve it in a fairly short time. Hodgkin was fortunate to be anywhere after the war. Part of his wartime job was testing radar equipment in aircraft during flight, and he was nearly killed when his airplane was hit by machine-gun fire from a British Spitfire whose pilot had mistaken it for a German plane. After the war in 1945, Adrian, who had been heading the Physiological Laboratory of Cambridge University, obtained Hodgkin's release from military service on the grounds that he was needed for teaching. Adrian had been responsible for running a large share of the teaching of medical students and nurses during the war. Hodgkin took responsibility for teaching the nurses and began to set up his laboratory in his spare time. He had loaned the electronic equipment he had used at Plymouth to a colleague, who had transported it back to Cambridge in 1940. This was lucky because Hodgkin's laboratory at Plymouth was destroyed in a bombing raid in 1942. Adrian obtained a five-year grant from the Rockefeller Foundation, which provided funding to support the salaries of Hodgkin and other promising researchers returning after the war. It is interesting that Hodgkin was hired at Cambridge for teaching but his salary support came from the Rockefeller Foundation in the United States for the purpose of supporting his research. Not just nongovernment organizations but, most important, the U.S. government was to enter into a period of unusually enlightened support of basic research after the war, which was further energized by the shocking launch of Sputnik into orbit by the Soviet Union in 1957.

Hodgkin's postwar research began in the fall of 1945. Fall and winter are not squid season, but in any case the Plymouth Laboratory's trawler, which fortunately survived the war, had not yet been given back by the military. In the meantime Hodgkin worked on single fibers of crab nerve using extracellular electrodes. Hodgkin was more a physicist by nature than a biologist, so he was interested in numbers, and he set about making a collection of some of the most interesting numbers associated with axons. He used a variety of approaches to get numbers from crab axons without intracellular electrodes, including immersing a crab axon in oil so that the extracellular ionic solution consisted only of the small amount of fluid clinging to the outside. Optical and electrical measurements gave Hodgkin a value of 3×10^{-6} milliliters of extracellular fluid surrounding a 1-centimeter (cm) length for an axon with a 30-micrometer (μm) diameter, immersed in oil.

This was not just a curiosity. The potassium in the thin film of fluid surrounding a 1-cm length of axon was just 3×10^{-11} moles. During his experiments in 1945, Hodgkin had discovered that increasing the extracellular potassium concentration increased the conductance of the membrane to potassium. We now know that this is because the depolarization caused by increased extracellular potassium opens the voltage-activated gates of more potassium channels, but Hodgkin had no idea why. Nonetheless, Hodgkin realized that he could estimate the extracellular potassium concentration in the thin layer surrounding an oil-immersed crab axon from its effect on the membrane potential. Stimulating a series of action potentials in an oil-immersed crab axon caused the resting membrane potential to depolarize, and the depolarization lasted for a considerable time after the stimulation had ceased, presumably because of the potassium released from inside the axon where the concentration is extremely high. In seawater-immersed axons there was no depolarization of the resting potential in response to a series of action potentials, because the released potassium diffuses away so that the concentration immediately surrounding the axon does not become significantly elevated. Assuming that the depolarization of oil-immersed axons was due to release of potassium by the axons collecting in the thin layer of surrounding extracellular fluid, Hodgkin (1947) estimated that 1.7×10^{-12} moles of potassium escaped through 1 cm² of membrane during the passage of each action potential. This amounts to a flow of about 10,000 potassium ions escaping across each square micrometer of membrane during each action potential. For comparison, a cubic micrometer of the cytoplasm inside the axon contains about 90 million potassium ions. Hodgkin calculated that the amount of potassium exiting the axon during a single action potential is equivalent to the flow of 0.17 microcoulombs of electric charge. This is about twice the charge difference across the membrane that produces the resting potential and enough to produce an overshooting action potential. Some years later, Richard Keynes joined the project and made direct measurements of the flow of radioactively labeled sodium and potassium across the membrane during action potentials, which confirmed Hodgkin's original estimates.

THE SODIUM THEORY

Bernard Katz, a physiologist and future Nobel Prize winner (see Chapter 7), joined the staff at University College London after the war. He performed experiments showing that crab axons could not conduct action potentials in salt-free sugar solutions (Katz 1947), which was in direct contradiction to the Curtis and Cole (1942) report of action potentials in squid axons deprived of all extracellular ions. This motivated Hodgkin and Huxley to test the sodium theory in crab axons during the winter of 1947, even though they could only estimate the membrane potential from the voltage at the cut end of the axon. They found that lowering the external sodium concentration to one-fifth lowered the amplitude of the action potential by 40%, which was close to the amount predicted by the Nernst potential for sodium. Thus, it looked like an increase in sodium permeability caused inward sodium current to become dominant during the action potential.

However, circumstances did not cooperate; a postwar coal shortage, which put an end to the heating of homes and buildings, along with a cold spell ended work on crab axons until spring. By then the Plymouth Laboratory was back in action, complete with its trawler, but there were still hardships. The laboratory was being rebuilt from the damage of the bombings in 1941, and Huxley was getting married and would be away for the summer. So Hodgkin began the summer alone, starting experiments in the late afternoon when the squids were brought in and working into the wee hours. Later in the summer he would collaborate with Bernard Katz and things improved.

Hodgkin used glass capillary electrodes inserted into the cut ends of the squid axons (see Chapter 4), and he quickly gathered evidence for the sodium theory. If inward sodium current produced by an increase in sodium conductance actually produces the rise of the action potential, then changing the sodium concentration in the external solution will change the sodium equilibrium potential and thus change the height of the peak of the action potential as he had shown with crab axons. Changing the internal sodium concentration in the axon would also work. That is much more difficult, although it would eventually become possible to squeeze the axoplasm out of a squid axon and replace it with a solution containing the desired sodium concentration. However, that only became feasible some time after Hodgkin and Huxley's experiments were completed. While the increase in sodium permeability during the action potential does not last long enough for the membrane potential to reach the sodium equilibrium potential, changing the sodium equilibrium potential by lowering or raising the external sodium concentration affects the height that the action potential does reach during the millisecond or so that the sodium conductance is increased.

Seawater is mostly sodium chloride with some potassium and traces of other ions. Hodgkin and Katz reduced the external ion concentrations by exposing squid axons to a mixture of seawater and isotonic dextrose. Dextrose, a sugar without electric charge, has no impact on the membrane potential, and the concentration of isotonic dextrose exactly matches the ion concentration in seawater so that the axon is not subjected to osmotic pressure. Hodgkin and Katz (1949) found that mixing various proportions of seawater and isotonic dextrose reduced the peak of the action potential by the amount predicted by the Nernst equation for sodium, which suggests that at the peak of the action potential the permeability of the membrane to sodium exceeded the permeability to all the other ions in seawater. In any biological system health is important, and a reduction produced by a treatment might mean that the treatment causes some deterioration, which could be responsible for the effect. One good control experiment is reversibility, showing that the reduction is reversed when the treatment is withdrawn. Another is to show an increase instead of a decrease, since increases cannot usually be ascribed to deterioration, which is what Hodgkin and Katz did. When squid axons were exposed to a solution containing 1.56 times the seawater concentration of sodium, the height of the action potential increased by an amount similar to that predicted by the Nernst equation.

Hodgkin and Katz also found that the depth of the undershoot after the falling phase of the action potential in squid giant axon depended upon the external potassium concentration, which is consistent with Hodgkin's discovery of the flow of potassium out of crab axons during the action potential described above. They realized that an accurate description of the permeability changes occurring during the action potential would require taking into account the permeability of sodium, potassium, and chloride using what would eventually be called the Goldman-Hodgkin-Katz equation (see Chapter 3). The data collected that summer allowed Hodgkin and Katz to estimate the changes in relative permeability that occur during the action potential. At the resting potential the relative permeability of the squid giant axon to potassium, sodium, and chloride is respectively 1:0.04:0.45. The sodium permeability is the least of the three, the potassium permeability is about 25 times the sodium permeability, and the chloride permeability is about 11 times the sodium permeability. The chloride permeability does not change during the action potential. At the peak of the action potential the sodium permeability increases to about 20 times the potassium permeability. During the undershoot, the sodium permeability falls to zero, and the potassium permeability is increased to almost double (1.8 times) its resting value. The major change that drives the action potential is, by far, the change in sodium permeability.

Hodgkin and Katz wrote two papers, one reporting this work and another reporting work on the effects of temperature on electrical activity in squid axon, which were published in the *Journal of Physiology* in 1949, more than 15 months after they had submitted the manuscripts. Being angry at journals is a constant; only the reasons vary. In his autobiography, Hodgkin (1992, 277) writes: "I was so fed up with these long delays, as well as being lectured at by a senior colleague on the dangers of over-publishing, that I temporarily deserted the *Journal of Physiology* and sent my next paper with Nastuk to the *Journal of Cellular and Comparative Physiology*."

THE VOLTAGE CLAMP

At the end of the 1947 squid season, Hodgkin wrote to Kacy Cole about his results with the squid axon and his plans for future research. One idea was to insert an electrode that was conductive over most of its length, rather than just at the tip, into the squid axon. Hodgkin thought that this would be a good way to study the action potential mechanism because a large stretch of axon membrane could be brought to threshold at once and would fire an action potential at once over its entire length, so that current spreading in local circuits and the propagation of a traveling action potential would not complicate the observations. In effect, the entire axon membrane would behave like a single location along the axon over which the action potential occurs simultaneously.

Cole wrote back informing Hodgkin that he had spent the 1947 squid season at Woods Hole doing just that, but combined with another major innovation. Because the membrane potential and the sodium and potassium conductance and currents all change during the action potential, it was necessary to hold something constant to perform an analysis. Cole and his collaborator George Marmount used a feedback circuit that held the membrane potential constant. Cole showed Hodgkin the voltage-clamp data during Hodgkin's visit to Cole's laboratory at the University of Chicago in the spring of 1948. The most interesting feature of the voltage-clamp data was that clamping the membrane at a depolarized voltage above the threshold for generating an action potential produced a brief inward current, followed by a longer-lasting outward current. We now know these are the sodium and potassium currents, respectively. The voltage clamp would provide Hodgkin and Huxley with the data they would use to discover the mechanism of the action potential. In his autobiography Hodgkin (1992) displays some embarrassment about using Cole's method to solve the action potential. He quotes a 1976 letter in which Andrew Huxley recalls that Hodgkin talked with him about a feedback system at least two years before he had visited Cole and saw the data produced by the voltage clamp. In any event, much time elapsed between seeing Cole's data and Hodgkin and Huxley's use of the technique, and almost five years elapsed before the publication of their analysis of the action potential, so it doesn't appear as if they rushed to scoop the analysis out from under a competitor.

The summer of 1948 was one of tinkering and tuning up their approaches to the squid axon. Hodgkin and Huxley developed a voltage-clamp electrode in which two bare wires oriented alongside one another could be extended along a considerable length of axon (Figure 5.1). The fine wires were wound in a double spiral around a fine glass rod, which provided support and prevented the wires from touching each other. The two wires were not connected in the same circuit. One internal electrode served to inject current into the axon to clamp the voltage at the desired level, and the other electrode was used to measure the membrane potential.

The long, bare-wire electrodes extending along a length of axon serve to short-circuit the internal resistance, so that the internal voltage is always the same within the entire length occupied by the electrode. Think of it this way: The external resistance is very low because of the large volume of solution in which the axon is immersed, and the internal resistance is made virtually zero by the insertion of the metal wires that short-circuit the resistance of the cytoplasm inside the axon. With the internal resistance virtually zero, the inside of the wired length of axon

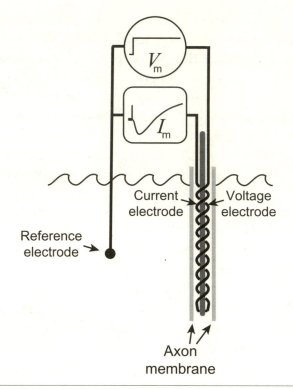

Figure 5.1. Helical voltage-clamp electrodes inserted into a squid giant axon.

will have a uniform voltage so that the entire membrane behaves electrically as a single location with no longitudinal current flowing along the axon. This arrangement had a definite advantage apart from the ability to voltage-clamp the membrane. If the membrane is not voltage-clamped and a depolarizing pulse is applied via the current injection wire, the entire membrane over the whole length is depolarized at once. If the depolarizing pulse reaches threshold, an action potential is fired simultaneously over the entire wired length of axon. Thus, the voltage profile along the wired length of axon is always flat; it rises up and falls back down everywhere in unison. Without the complications of longitudinal currents flowing through the internal resistance of the axon, the membrane potential in the entire wired length of axon can be modeled by the simple circuit shown in Figure 5.2 (see Chapter 3 for circuit diagram details). This is the same circuit depicted in Figure 3.11.

The voltage-clamp circuit automatically injects exactly the amount of current into the axon needed to oppose any capacitative current

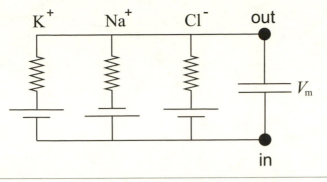

Figure 5.2. Circuit model of a single location along the squid axon membrane.

flow across the membrane. Since there is no capacitative current flow, the voltage clamp holds the membrane potential at the voltage selected by the experimenter. In a voltage-clamped axon, all current flow across the membrane is ionic current, and the amount of current transferred across the membrane through the electronic feedback circuit to hold the voltage constant is exactly opposite to the combined ionic currents driven through the membrane by the potassium, sodium, and chloride batteries. Thus, the voltage clamp provides a means to determine the ionic current flowing across the membrane.

Figure 5.3 depicts the pattern of current flow across a squid giant axon membrane when the membrane potential is clamped at a voltage depolarized above the threshold for firing an action potential. The voltage clamp is turned on at 0 milliseconds (msec). When the amplitude of the trace is below 0 milliamps (mA) the current is inward across the membrane, and when the trace is above 0 mA the current is outward across the membrane. Even though the depolarization is above threshold, no action potential is fired because the feedback circuit holds the membrane at the clamped voltage.

When the depolarizing voltage is turned on at time = 0, there is a brief spike of outward capacitative current, which establishes the clamped voltage. After depolarizing the membrane, the capacitative current quickly goes back to zero because the membrane potential is held constant by the voltage-clamp circuit. Because the capacitative current is zero, all subsequent current across the membrane is ionic current. The depolarization causes the membrane to undergo changes in ionic permeability that produce a net inward ionic current lasting about 1.5 msec, shown dipping below the dotted line in Figure 5.3, followed by a

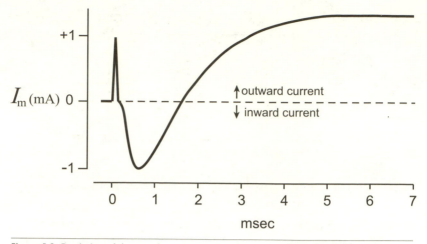

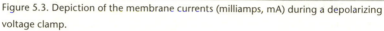

Figure 5.3. Depiction of the membrane currents (milliamps, mA) during a depolarizing voltage clamp.

net outward current rising above the dotted line. The outward current persists as long as the voltage clamp is maintained. As described in Chapter 3, we now know that the inward current is sodium current, which turns on because depolarization opens the fast sodium activation gates and then shuts off automatically because depolarization also closes the slow sodium inactivation gates. The outward current then rises because depolarization opens more of the slow potassium activation gates, which increases the number of open potassium channels. Potassium channels do not have inactivation gates, so the additional potassium channels remain open as long as the clamped voltage is maintained. How much current crosses the membrane depends upon the magnitude of the depolarizing voltage; the greater the depolarization, the greater the inward and outward currents because more sodium and potassium channels are opened.

The inward sodium current and outward potassium current are net currents, and all ions carry currents when the membrane potential is not at their equilibrium potential (see Figure 3.10), so the sodium and potassium currents overlap in time. Hodgkin and Huxley employed a clever maneuver in order to determine the actual values of the sodium and potassium currents separately. First, they clamped the membrane at a depolarized voltage and recorded the combined inward and outward currents that followed. Next, by replacing just the right amount of the sodium in the bathing solution with choline, a positive ion that cannot cross the membrane, they reduced the sodium equilibrium potential to exactly the voltage at which the membrane had been clamped. Thus, when the clamped voltage was turned on again, the sodium current was zero, and nothing happened until the slow potassium gates opened, producing the rise of outward potassium current in isolation. In this way they determined the potassium current. (Clever, huh!) Subtracting the potassium current from the previously determined combined currents produced the sodium current. Since potassium and sodium currents could be determined, and the voltage across the membrane was known, the potassium and sodium conductance at any time during the voltage clamp could be calculated by Ohm's law: $g_K = I_K \div (V_m - V_K)$ and $g_{Na} = I_{Na} \div (V_m - V_{Na})$.

These measurements were made at a variety of clamped voltages and external sodium concentrations, and the family of curves so obtained provided Hodgkin and Huxley with the relationships between membrane potential and sodium and potassium conductance they used to derive the differential equations that model the action potential (Hodgkin 1992). Hodgkin and Huxley's analysis showed that inward sodium current produced by depolarization ignites a positive feedback cycle that produces the action potential: depolarization of the membrane causes an increase in sodium permeability, which causes the influx of sodium, which in turn causes further depolarization of the membrane, which further increases the sodium permeability, which causes the influx of more sodium, and so forth until the cycle is interrupted by the closure of the sodium inactivation gates. The voltage-clamp experiments also provided the data needed to reconstruct a traveling action potential as it moves along the membrane (see below).

Hodgkin and Huxley performed all the experiments for their landmark papers (Hodgkin and Huxley 1952a, 1952b, 1952c, 1952d) between mid-July and mid-August 1949. The complex voltage-clamp data took over two years to analyze. The Cambridge University computer was out of action for a critical six months while getting an upgrade, which did not help. Laptop computers, desktop computers, and even the Texas Instruments handheld electronic calculator that replaced the mechanical calculating machine and the slide rule were yet to come. Andrew Huxley solved the equations that describe the propagation of the action potential during three grueling weeks bent over a mechanical calculator.

The curves in Figure 5.4 provide a dynamic picture of the major players in the action potential mechanism—the membrane potential, the sodium conductance, and the potassium conductance—as they travel together along the axon. We can see that the beginning of the passive rising phase of the action potential (a) to threshold (b) precedes the increase in sodium conductance. The passive rise is caused by currents spreading forward from the region of active rise of the action potential (b–c) where the depolarization is driven by the fast-rising inward sodium current through the sodium channels whose activation gates have been opened by the depolarization. The action potential peaks and the falling phase of the action potential (c–d) is produced by a combination of the closure of the sodium inactivation gates along with the opening of more potassium activation gates, both of which occur as a delayed response to the depolarization of the active rising phase. Then, the action potential undershoots the resting potential (d–e) on the tail of the increased potassium conductance, which then falls to the pre-action potential level, restoring the resting potential (e).

Hodgkin and Huxley had their disappointments, and they did not regard the project as 100% successful. One of their major goals had been to discover the physical basis for the changes in sodium and potassium permeability. Basically, there are two conceivable ways that ions can cross the hydrophobic cell membrane: riding in vehicles protected from the hydrophobic environment that shuttle back and forth across the membrane or passing through channels lined with electric charges. In membrane parlance, vehicles are called carriers. Hodgkin and Huxley's pet theory at the beginning of their experiments was that sodium and potassium crossed the membrane riding in carriers. They soon realized that carriers shuttling ions back and forth across the membrane would not likely be able to transfer ions across the membrane fast enough to account for the rise in current that occurred when a depolarizing voltage was applied to the membrane. In the words of Alan Hodgkin (1992, 291): "We soon realized that the carrier model could not be made to fit certain results, for instance the nearly linear instantaneous current-voltage relationship, and that it had to be replaced by some kind of voltage dependent gating system. As soon as we began to think about molecular mechanisms, it became clear that the electrical

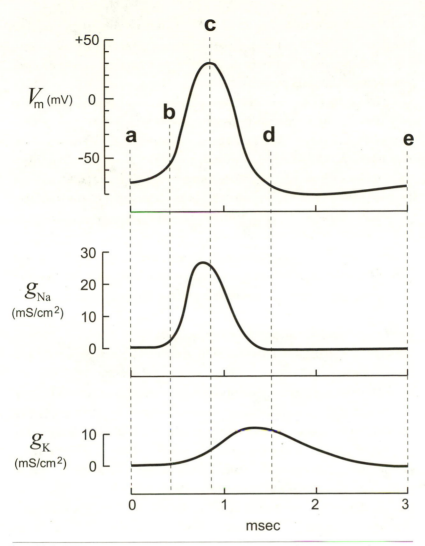

Figure 5.4. Depiction of the changes in sodium conductance and potassium conductance (millisiemens, mS) during the action potential.

data would by itself yield only very general information about the class of system likely to be involved. So we settled for the more limited aim of finding a simple set of mathematical equations, which might plausibly represent the movement of electrically charged gating particles. But even this was not easy."

Hodgkin and Huxley's insight that the changes in ionic permeability are controlled by the opening and closing of voltage-dependent gates on ion channels was prophetic indeed. Just as knowledge of electricity had to catch up to Galvani's insight that nerve impulses are electrical, knowledge of cells, membranes, and proteins had to catch up to Hodgkin and Huxley's insight into how membrane channels must work to produce the action potential.

The biological proof of the action potential mechanism came pouring in from many directions after the Hodgkin and Huxley analysis. The nervous system is the target of many poisons in nature, and the discovery and use of a wide range of poisons has led to much information about how the nervous system works. It is a bad idea to eat Japanese puffer fish. Puffer fish gonads contain one of the most potent poisons in existence. I cannot imagine why anyone who knows this would eat a puffer fish, but there are restaurants in Japan that are licensed to serve it to their patrons. The chefs are trained to clean puffer fish without contaminating the muscle with gonadal tissue. Delicacy or not—no thank you! The poison is tetrodotoxin; it binds to the outside end of sodium channels and prevents sodium from going through—and you need your sodium channels very much.

Another drug, tetraethylammonium (TEA, for short), similarly blocks potassium channels. The use of tetrodotoxin and TEA on squid axons relatively effortlessly revealed the individual sodium and potassium currents that Hodgkin and Huxley had derived from their experimental analysis. These channel blockers have been used to estimate the density of sodium and potassium channels in axons, which is fairly sparse, about 300 per square micrometer for sodium channels and 70 per square micrometer for potassium channels in squid giant axon. Nodes of Ranvier, the tiny lengths of bare axon between successive myelin sheaths (see below), in frogs have about 10 times that density of sodium channels in their membranes.

In 1949 the funding of the Rockefeller Unit was ending, and Adrian tried to get Cambridge University to take over its funding, including Hodgkin's salary, but this was not successful. Hodgkin, who had completed the experiments that would earn him and Andrew Huxley the Nobel Prize in 1963, was not given a permanent faculty position. Adrian arranged for Hodgkin to stay on as assistant director of research, a position that had to be renewed every five years and was not well paid. In 1951 the situation was corrected when Hodgkin was appointed to the Foulerton Research Professorship of the Royal Society on Adrian's recommendation.

Many scientific achievements look in retrospect like a sequence of well-planned experiments that yield the predicted outcomes, which nicely fall into place to support the investigator's hypothesis. This is rare. In his autobiography Hodgkin comments: "Fitting theoretical equations to biological processes is not always particularly helpful, but in this case Huxley and I had a strong reason for carrying out such an analysis. A nerve fibre undergoes all sorts of complicated electrical changes under different experimental conditions and it is not obvious that these can be explained by relatively simple permeability changes of the kind [we have described]. To answer such questions we needed a theory and preferably one that could be given a physical basis of some kind" (1992, 297).

Hodgkin's misgivings notwithstanding, their analysis did work, continues to stand the test of time half a century later, and has been shown to have wide application throughout the animal kingdom. Hodgkin and Huxley had chosen their problem well. It has turned out that animal electricity is electricity; it is not some biochemical reaction that gives off electricity as an epiphenomenal by-product. As electricity it follows the rules of electricity—it has no choice—so the electrical activity of the squid axon met the demands of their equations with astounding accuracy. While they employed techniques first developed by others, their vision seems to have been uncannily clear when they undertook their experiments in Plymouth during the summer of 1949. All the tools were available, but no one else seems to have even come close.

THE TRAVELING ACTION POTENTIAL

As with any phenomenon where many things are happening simultaneously, understanding how the action potential travels along the axon requires some imagination to picture the outcome of all those processes happening at once. For a tiny fraction of humans, mathematics is enough, so differential equations such as those developed by Hodgkin and Huxley provide the key to understanding—some believe the only real key. Richard Feynman said this isn't so: "if we have a way of knowing what should happen in given circumstances without actually solving the equations, then we 'understand' the equations, as applied to these circumstances. A physical understanding is a completely unmathematical, imprecise, and inexact thing, but absolutely necessary for a physicist" (Feynman, Leighton, and Sands 1964, 2.1).

I had a personal experience of what Feynman meant in the first years of my career as an assistant professor. I was serving as the token biologist on the Ph.D. supervisory committee of a student in applied physics whose thesis involved electrical studies of ion channels in cultured cells. During the student's Ph.D. candidacy exam I asked some questions about what effect on the action potential would he expect if a certain parameter that I specified were changed. The format of my questions was: "If X increased, would Y increase or decrease?" Anyone who has what I would call a real understanding of the action potential would be able to answer these questions off the top of his head without much fuss. Instead, he said something like "Well ... let me see ... ," wrote some equations on the board, and then gave the correct answer. Just like a computer, which functions without understanding what it is doing, the student was able to come up with the correct answer every time. This student was extremely bright and, in fairness, under the pressure of the exam may have just been relying upon the most trustworthy standard of truth that he knew. But the story still makes my point that an intuitive understanding of the processes operating in our world is a valuable thing worth cultivating because we rarely—most of us never—solve the equations or even know what the equations are. This and other experiences teaching caused me to develop a mildly sadistic stock question that I have asked all students during their Ph.D. oral exam: "Can you tell me briefly what you have discovered in your thesis work without indicating the specific animal or cell that you worked with or the specific experiments you performed?" The idea is to determine if the student has a grasp of the generality and importance, if any, of their discoveries rather than just the ability to rattle off exactly the results of their experiments. Many students just stared at me blankly, apparently with no idea what I meant. One wonders whether they did all those experiments without really understanding why. These days many students function as cogs in a research machine, and even some of those aiming for the highest degree that academia has to offer may not know the place of their work in the great scheme of things. When one's area of specialty is tiny, as is the trend, understanding of the world shrinks. There is a pejorative description of expertise that applies:

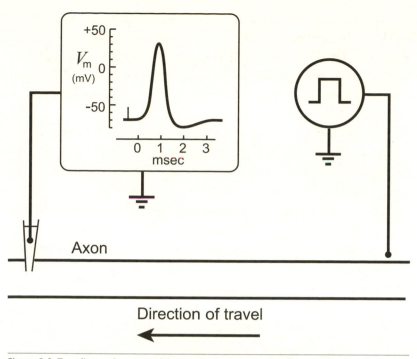

Figure 5.5. Traveling action potentials recorded passing by an intracellular microelectrode.

knowing more and more about less and less. No one can know the mathematics underlying every field, so an intuitive approach is the only way to achieve a personal understanding of the broader world.

Now let's complete our intuitive picture of the action potential by examining how it travels along an axon, starting with what an action potential really looks like. In order to do this we need to do some thought experiments where traveling action potentials are observed as a change in voltage across the membrane, measured as they pass by a recording electrode (Figure 5.5). Traveling action potentials can be measured in squid axon with glass capillary electrodes inserted into the cut end of the axon that only detect the voltage at their tips. Glass microelectrodes that can be inserted across the cell membrane were developed about the same time that Hodgkin and Huxley were initiating their voltage-clamp experiments (see Chapter 4). These electrodes have been widely used for measuring the membrane potentials of neuronal cell bodies and muscle fibers in studies of synaptic transmission (see Chapter 7). However mammalian axons are too small to tolerate the insertion of glass microelectrodes, so the following observations are thought experiments that, although they cannot be performed using most axons, are based upon knowledge of the action potential obtained by Hodgkin and Huxley during their voltage-clamp experiments and corroborated in many other ways.

The intracellular microelectrode depicted in Figure 5.5 records the voltage profile of the action potential as it passes by. All action potentials in nerve rise to a peak in about 1 msec and are over in about 3 msec (Figure 5.6a). If an action potential is traveling at a velocity of 1 meter per second (m/sec), which is equivalent to 1 mm/msec, then the leading edge of the action potential voltage profile precedes the peak by 1 mm and precedes the tail end of the profile by 3 mm. This conversion from milliseconds to millimeters works for any point along the action potential. Therefore, at the conduction velocity of 1 mm/msec, one can plot the voltage profile of the action potential along the axon by simply converting the millisecond timescale to a millimeter distance scale, and the shape looks just the same (Figure 5.6b). This conversion amounts to just multiplying the values in the timescale by the conduction velocity, which yields the distance the action potential voltage profile must extend along the axon to produce an action potential that lasts 3 msec as it passes by the recording electrode.

All action potentials in all nerves last about 3 msec because the voltage-gated sodium and potassium channels in all axons have similar kinetics. Therefore, action potentials in fast-conducting axons must have longer voltage profiles than action potentials in slow-conducting axons as they travel along. In order to appreciate the full impact of this, consider the experimental setup in Figure 5.5 where an axon is being stimulated at one location to produce action potentials, which are being recorded at another location some distance away. For the purposes of the illustration let's stimulate the axon at a rate of about 333 per second such that an action potential is produced every 3 msec, and there is no space between action potentials as they travel along the axon.

Figure 5.7a shows the voltage profile along a 300-mm stretch of axon between the stimulating electrode and the recording electrode when the conduction velocity is 1 m/sec, which is characteristic of the slow-

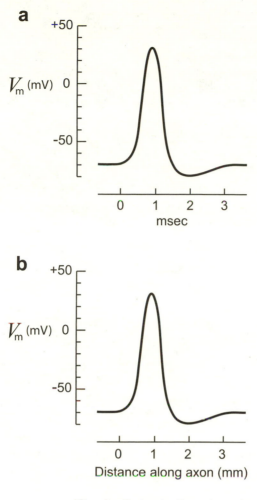

Figure 5.6. Depictions of (a) an action potential traveling at 1 mm/msec as it passes by the recording electrode; and (b) the voltage profile of the action potential as it travels along the axon.

conducting axons that conduct pain signals in mammals. This is the velocity of a slow walk (2.24 miles per hour). Since action potentials are 3 mm long at this velocity, there are 100 action potentials lined up along a 300-mm stretch of axon on their way to the recording electrode. If the conduction velocity is increased to 10 m/sec (Figure 5.7b), the voltage profile of each action potential is expanded 10-fold to 30 mm of axon,

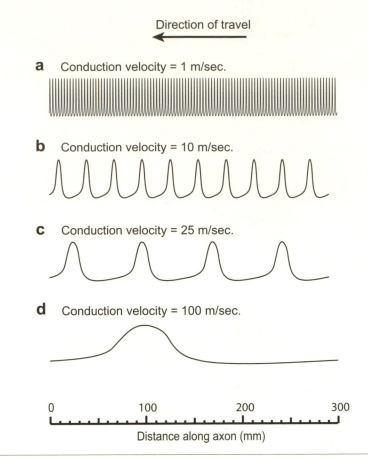

Figure 5.7. Voltage profiles depicting action potentials firing at a rate of 333/sec, traveling in axons with different conduction velocities.

so there are only 10 action potential profiles lined up along a 300-mm stretch of axon.

Squid giant axons ramp the speed up a bit more than that, conducting action potentials at 25 m/sec (Figure 5.7c), which is moderate motor vehicle speed (56 miles per hour). At this velocity the action potential duration-to-length conversion is 25 mm/msec, and each action potential extends along 75 mm of axon so only four are accommodated along a 300-mm stretch of axon. The conduction velocity of myelinated axons innervating skeletal muscle in mammals is about 100 m/sec (d), which is moderate aircraft speed (224 miles per hour), so just a single action potential occupies an entire 300-mm stretch of axon on the way to the recording electrode. How myelinated axons of vertebrates, which have diameters only a fraction of the diameter of squid giant axons, achieve a conduction velocity four times greater is discussed below.

Now let's assemble the electrical events underlying the action potential with what the action potential profile looks like as it travels along the axon. A portion of an action potential profile is depicted in Figure 5.8. The axon occupied by the action potential is divided into the region of the passive rising phase, the active rising phase, and the falling phase. The undershoot produced by the residual increase in potassium conductance does not play a major mechanistic role and is not depicted. Each of the three regions is modeled in the corresponding circuit diagram. Potassium, sodium, and chloride channels are interspersed all along the axon, so a complete circuit diagram would have sodium, potassium, and chloride channels in all three regions, but including them all would create an incomprehensible mess. Instead I have only depicted the ion channels that produce the net ionic current in each region.

In the region of active rise of the action potential the sodium conductance rises steeply, resulting in the inward flow of sodium current driven by the sodium battery (labeled a in Figure 5.8). The inward sodium current drives outward current across the membrane capacitance in this region (b), depolarizing the membrane potential to reach the crest of the action potential. A component of the inward current travels forward inside the axon (c) and then departs as outward current across the membrane capacitance (d), producing the depolarization of the passive rising phase at the leading edge of the action potential. The passive depolarization reaches threshold at the border with the active rising phase. This forward current loop from the region of the active rising phase into the region of the passive rising phase of the action potential is the local circuit that Alan Hodgkin discovered that crosses a cold-blocked region of the axon (Figure 4.5).

The closure of the sodium inactivation gates and the opening of more potassium activation gates cause the action potential to crest and then enter the region of the falling phase of the action potential. With the sodium conductance turning off and the potassium conductance increasing, the potassium and chloride equilibrium potentials regain

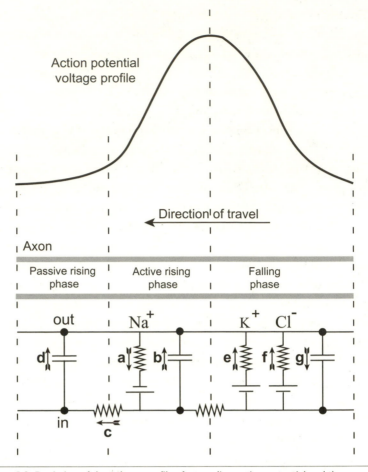

Figure 5.8. Depiction of the voltage profile of a traveling action potential and the corresponding circuit model.

their roles as the major determinants of the membrane potential. The result is that a net outward current through the potassium channels (Figure 5.8e) and chloride channels (f) drives inward current across the membrane capacitance (g) that repolarizes the membrane potential in the region of the falling phase. The falling phase of the action potential is as important as the rising phase. Without a falling phase, transmission would end after the passage of the first action potential—think of a dynamite fuse, which can only burn once.

The forward longitudinal current that depolarizes the membrane to threshold is not the only current flowing inside the axon during the action potential. Ohm's law tells us that current flows between any electrically connected points that have different voltages. A glance at the action potential voltage profile in Figure 5.8 immediately tells us that the crest of the action potential where the membrane potential is most positive is like a voltage watershed because the membrane potential is progressively less positive along the axon in both directions, forward and backward. This means that in addition to the forward current loop, there is also a current loop projecting backward from the crest of the action potential and crossing outward across the membrane along the trailing edge of the action potential. Thus, there is symmetry here with the current entering the axon projecting forward and backward and then leaving as outward current across the axon membrane. The asymmetry that gives the action potential a front and a back and a direction of travel is that the outward current driven by the forward current loop leaves the axon as capacitive current that depolarizes the membrane, while the outward current driven by the backward current loop leaves through the potassium and chloride channels, which contributes to the repolarization of the membrane.

There is a time interval after the crest of the action potential when enough sodium inactivation gates are closed to prevent any passive depolarization from firing another action potential. This period when the membrane is incapable of producing another action potential is called the refractory period. Action potentials can travel in either direction along an axon. Their normal directionality arises because in the vast majority of neurons action potentials are generated at the base of the axon where it joins the cell body, called the initial segment of the axon, and travel from the cell body toward the axon terminals. If action potentials traveling toward each other are artificially generated by stimulating an axon simultaneously at two locations, the action potentials destroy each other upon colliding because their leading edges encounter the refractory regions where the oncoming action potential has closed the inactivation gates of the sodium channels.

While the circuit in Figure 5.8 is a useful tool for understanding how action potentials travel along the axon, the membrane potential and sodium and potassium conductance are smoothly changing along the axon membrane within the three regions, which is not captured in the circuit diagram. An axon divided into three discrete sections, within each of which the membrane ionic conductance, membrane current, and membrane potential are uniform, would display a very strange action potential that resembles a square wave, but that is, of course, not what happens. During the smooth rise and fall of the action potential, current loops project from every infinitesimal location as the action potential rises to its crest and falls back to the resting potential. Current loops project forward from more positive regions into less positive regions all along the region of the active rise of the action potential, and those extending into the region of passive rise all contribute to bringing the membrane ahead of the active rising phase to threshold and moving the action potential forward.

How much current extends forward and backward throughout the action potential profile as it travels along depends upon the internal resistance and membrane capacitance per unit length of axon. The electrical resistance inside a given length of axon is inversely proportional to the cross-sectional area for the same reason that the electrical resistance of a copper wire is inversely proportional to its cross-sectional area; double the cross-sectional area and twice as many carriers of electric current—electrons in wires, ions in axons—are available to carry the current. Cross-sectional area increases with the square of the diameter—double the diameter, the cross-sectional area increases fourfold, and so the electrical resistance per unit length of axon decreases fourfold. The capacitance of the membrane of a given length of axon depends upon the surface area—double the diameter, the surface area doubles, and the capacitance per unit length of axon doubles.

In a larger-diameter axon, the decreased internal resistance causes the current loops to extend farther along the axon. The increase in membrane capacitance per unit length has the opposite effect because it allows more current to cross the membrane per unit length of axon, leaving less current to extend farther along the axon. However, this effect is mitigated because a larger surface area also means more sodium and potassium channels, which generate more current. Since the reduction in internal resistance with increasing diameter is greater than the increase in capacitance, the reduction in resistance dominates as axon diameter becomes larger, and so the current loops generated in large-diameter axons spread farther along the axon. The increased spread of current along larger axons occurs everywhere throughout the region of the axon occupied by the action potential, which makes the action potential profile occupy a longer stretch of axon as it travels along. Therefore, action potentials occupy a longer length of larger-diameter axons.

The kinetics of the opening and closing of channel gates is unaffected by axon diameter. Once the action potential has been initiated at any point along the axon, the action potential will go through its paces and be completed at that point (that is, will have passed that point) in about 3 msec. This means that each point along an action potential stretching out over a longer length of a larger-diameter axon still must go through the cycle of conductance changes to sodium and potassium within 3 msec, just like an action potential that only occupies a short length of a small-diameter axon. Thus, an action potential stretching over a longer length of a larger-diameter axon must travel faster to completely pass each point along the axon in only 3 msec; action potential size and velocity are inextricably linked together. Think of a mouse and a lion running past a tree on the African savanna. Suppose that from the nose to the tip of the tail the lion is 20 times the length of a mouse. If they both have to pass by the tree in the same amount of time, the lion must run 20 times faster than the mouse. (I would guess that lions have the muscle and the stride to do this. Although maybe not. Have you ever tried to catch a mouse?) This is a fundamental insight about the action potential—the reason that larger axons conduct action potentials faster. It is the reason squids have the giant axons that enabled Hodgkin and Huxley to figure out the mechanism of the action potential. This whole story was made possible because J. Z. Young, a zoologist, discovered giant axons in squids. Although most scientists, including zoologists, do not discover something with this level of impact, there is just no substitute for the curiosity-driven search for knowledge of how the world and the creatures within it work.

THE STRANGE CASE OF THE JUMPING ACTION POTENTIAL

As fast as they are, squid giant axons only conduct action potentials at about 25 m/sec. Mammalian axons of much smaller diameter conduct at velocities four times greater because of their myelin sheaths. Myelin is a fatty sheath of insulation produced by the winding of many layers of cell membrane originating from cells positioned along the axon. The myelin-making cells, Schwann cells in the peripheral nervous

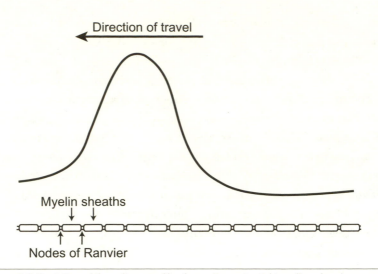

Figure 5.9. Depiction of the voltage profile of an action potential traveling along a myelinated axon.

system and oligodendrocytes in the brain and spinal cord, produce large broad sheets that consist of cell membranes pinched together with the inside surfaces touching each other and just about no intracellular space left at all. This sheet, two cell membranes thick, wraps around a section of axon many times, like wax paper on the roll, forming a thick, electrically insulating layer. The sections of myelin sheath are just a millimeter or so long, and the axon is myelinated by multiple sections of sheath, arranged one after the other, each separated from the next by a 1-micrometer-long gap called the node of Ranvier, where uninsulated axon membrane is exposed to the extracellular fluid (Figure 5.9). The insulated sections of axon between nodes of Ranvier are called internodes. (Note that in the figure the sections of myelin sheath are not to scale with the action potential; action potentials of the fastest-conducting axons span across hundreds of internodes.) Little, if any, current crosses the axon membrane in the internodes because the thick insulating layers of myelin greatly decrease the membrane capacitance by increasing the gap between the axoplasm and the extracellular fluid. This reduces any capacitative current to near zero, and there is little or no sodium or potassium current crossing the membrane because the intermodal axon membranes are relatively

free of sodium and potassium channels, the vast majority being localized at high concentration at the nodes of Ranvier.

Alan Hodgkin (1963) used a metaphor of a jumping action potential in his Nobel lecture to tell how myelin increases conduction velocity:
"You may wonder how it is that we get along without giant nerve fibres.
The answer is that vertebrates have developed myelinated axons in which
the fibre is covered with a relatively thick insulating layer over most of
its length, and the excitable membrane is exposed only at the nodes of
Ranvier. In these fibres, conduction is saltatory and the impulse skips
from one node to the next." The metaphor of action potentials skipping
or jumping from one node of Ranvier to the next, thereby achieving a
great increase in conduction velocity, has been neuroscience textbook
dogma for over half a century, and *saltatory conduction* became an accepted term. When I first encountered it as a student, I believed it, of
course, but I never felt that I really understood it—that was true for many
features of the action potential at the time. Years later while preparing
a lecture, I was startled by the realization that since the nodes of
Ranvier are spaced about every millimeter along the axon, the voltage
profile of a 300-mm-long action potential traveling at 100 m/sec along a
myelinated axon must span about 300 nodes of Ranvier. How could an
action potential of this physical length be described as hopping, skipping, or jumping between nodes of Ranvier? In the middle of the night
before my lecture this metaphor began to conjure up images of elephants
crossing a 1-meter-wide pond by tiptoeing from one lily pad to the
next—the elephant would not even fit in the pond, never mind on a single
lily pad. The action potential profile cannot exist at one node of Ranvier and hop to the next any more than an elephant could stand on one
lily pad and hop to the next like a frog could. But who was I to question
dogma coming down from the highest pinnacles of the field? I was near
panic, as I conjured up the fear that a student might notice this too and
ask a question. Such are the irrational terrors of a beginning university
teacher.

I have since come to understand how myelin works. The sole function of the internodes is to conduct current within the axon from one
node of Ranvier to the next. Because of the insulation that keeps the current inside the internodal axon, the flow of longitudinal current along
the internodes is like the flow of current in a wire. Consider the voltage

profile of the rising phase of the action potential, 1 msec in duration and 100 mm long in a fast-conducting myelinated axon. There are 100 nodes of Ranvier distributed along the rising slope of the action potential profile. The membrane potential at each node is more positive than the membrane potential of the node ahead, and current flows forward within the axon from one node to the next all along the rising slope of the action potential. The electric field produced by the voltage difference between each node and the node ahead travels forward through the axon at near light speed just like the electric field in a circuit of insulated wires when the switch is turned on. This electric field produces an immediate shift in the ions all along the internode, which immediately produces flow of current into the node ahead; that is, just as in a wire, charge does not have to travel all the way from the node behind before charge starts entering the node ahead. The only process that takes any time on the biological scale is the time required for enough charge to accumulate at each node to depolarize it to threshold. The higher the internal resistance of the internodes the smaller the longitudinal current, and the longer it takes for the depolarization of the rising phase of the action potential to move forward. However, at an internodal length of just 1 mm, the time required for the forward spread of the depolarization to reach threshold is nowhere near the time required for an action potential to travel along an equivalent length of unmyelinated axon, which requires activating the sodium and potassium channels all along the entire membrane.

Think of each node of Ranvier as a power station and the internodes as wires connecting them. Inward current generated at each node of Ranvier spreads forward through the internode and into the node ahead. Because of the internal resistance of the internode, the current projecting to the node ahead is just a little bit less than it would be if the two nodes were in direct contact with each other with no internode in between. Thus, the depolarization of the node ahead in the myelinated axon, spanning a distance of 1 mm, occurs almost as fast as an equivalent section of unmyelinated axon depolarizes the next section with no internode in between, but the additional millimeter makes the distance traversed many times longer and thus the conduction velocity many times faster.

Myelin offers another example of how the size of the action potential and velocity are inextricably linked. Action potentials in unmyelinated and myelinated axons have a similar time course because the rate-limiting steps are the opening and closing of the sodium and potassium channel gates, which have similar kinetics in all axons. In myelinated axons, connecting the nodes with fast-conducting internodes broadens the spread of the longitudinal currents over a longer length of axon without requiring much additional time. Consequently this broader action potential, rising and falling with nearly the same kinetics, covers a much greater distance in the same amount of time—the elephant does not hop but it must run faster than a cheetah to pass by a point along the axon in the same 3 msec that a smaller action potential passes by at a walk. You might wonder how a series of action potentials racing close together at 300 per second along a myelinated axon avoid running into each other when they must slow down upon entering a terminal branch that is smaller in diameter and lacks myelin. This is no problem at all, since slow action potentials have correspondingly shorter voltage profiles. Thus, when an action potential slows down upon reaching unmyelinated terminal branches, the action potential voltage profiles always shrink in size so that they just fit. In this way, the frequency of the signal traveling rapidly along a myelinated axon is preserved when the action potentials reach the axon terminals.

Finally, the action potential does not jump from node to node because the entire length of the myelinated axon occupied by an action potential profile at an instant in time, internodes included, participates in the depolarization and repolarization, which is the action potential. A slightly better phrase used by some neuroscientists is that in myelinated axons "excitation" jumps from node to node, but I do not like that either. The internode is like a wire, which transmits an electrical signal between nodes at near light speed, and saying that the signal jumps over the internode is like saying that a telephone conversation jumps over the landline connecting one telephone to the next. There is just no such thing as a jumping action potential.

The action potential is the animal electricity that Galvani and many others were looking for. It bears no resemblance to animal spirits flowing in tubular nerves. Galen and the other proponents of the animal spirit theory over the millennia had no idea of the tremendous amount of basic scientific knowledge that would have to accumulate before conduction in nerve and muscle and all their manifestations right up to consciousness and thinking could become subjects of human understanding.

6

HEART TO HEART

Until the late nineteenth century, no one knew what neurons really are. The realization that neurons are individual cells, not an electrically conductive network, raised the issue of how neurons communicate with each other. The realization that sensory cells communicate with neurons and neurons communicate with each other and with muscle by releasing chemicals came from a single experiment with frog hearts that required no high technology. All that was required was the ability to stimulate a nerve innervating the heart and count the beats. The main ingredients were the imagination to dream up the experiment, the luck to have it work, and the insight to realize that the results had broad implications way beyond the heart.

The story begins with Santiago Ramón y Cajal (1852–1934), who was the son of a physician in a rural backwater hidden within the scientific backwater that was Spain (see Finger 2000). Ramón was his father's surname, which in some of his writings he dropped in favor of just Cajal, his mother's. The reason could have been his stormy relationship with his father while growing up. The young Cajal was a nonconformist, had poor grades in school and many behavior problems, including some stealing and vandalism (blowing up a neighbor's gate with a homemade cannon), and wanted to be an artist—not a formula for a good relationship with a doctor father. But much of that changed when his father took the teenaged Cajal to a cemetery to introduce him to the wonders of human anatomy. They didn't have to dig anyone up because the cemetery owners exhumed bodies if the leases on the burial plots were not renewed. Cajal was enthralled, spent many happy hours drawing the bones of the evicted corpses, and made his father happy by deciding to become a doctor. His career took numerous twists and turns before he settled into his studies of the cellular anatomy of the nervous system. After graduating from the medical school in Zaragoza, Spain, he served a stint as a doctor with the Spanish army in Cuba, where he almost died of malaria before finally being granted a medical discharge. After arriving home in Spain he held professorships in Valencia and Barcelona. Early in his career he directed some of his attention to studies of cholera during an epidemic, for which the Spanish government rewarded him with the gift of a Zeiss microscope. That microscope became his constant companion while he pursued the work that changed the science of the nervous system.

Some years earlier Camillo Golgi (1843–1946), then a 30-year-old physician tucked away in his own backwater at Abbiategrasso hospital in Italy, began experimenting with silver stains in one of the hospital's kitchens. Silver stains, which at the time were in use in photographic development, seemed worth a try. In one sense the silver nitrate stain developed by Golgi was a failure since only about 3% of neurons in a section (a very thin slice) of the brain were stained. However, the neurons that were stained looked like they were stained completely throughout their entire anatomy. To this day it is not known what difference causes one neuron to fully stain and another not at all. Since the cell bodies and dendrites of neurons are densely packed together in the brain, and since axons run together within fiber tracts, if all or even a substantial fraction of the neurons were stained, most of the stained neurons would have overlaid one another, making it impossible to tell where one neuron ends and the other begins. So the Golgi method, which only stains a tiny fraction of the neurons, some of which are few and far between, allowed for the first time the study of the detailed anatomy of many of the different kinds of neurons in the brain and spinal cord.

Golgi had published some of this work in 1873, but information flow was not what it is today. In 1887 Cajal was shown nerve tissue stained with the Golgi method while visiting Luis Simarro, a psychiatrist at the University of Madrid who had a far better equipped laboratory than his own. He immediately recognized the Golgi method as far superior to anything he had seen before. This was the turning point in Cajal's lifework. Scientific interaction was not what it is now either, and Cajal and Golgi would not actually meet until 1906 in Stockholm when receiving their shared Nobel Prize in Medicine or Physiology. Golgi's Nobel Prize lecture, which he delivered first, followed by Cajal, was laced with a diatribe against Cajal's neuron doctrine (see below). As it turned out, they did not hit it off. Nowadays the powers that be tell us that solitary geniuses in science, if they ever really existed at all, are certainly a thing of the past, and creative discoveries come only through collaboration. However, not all collaborations are made in heaven, and I wonder if Cajal would have made anywhere near the major discoveries credited to him if he had entered into collaboration with the more senior Golgi early in his career.

In the latter part of the nineteenth century the science of the nervous system was in a period of figuring out what needed to be figured out. The lay of the land was established by the defining of a few important terms: *neuron* in 1891, *dendrite* in 1890, *axon* in 1896. In 1897 Charles Sherrington described axon terminals as forming connections he called "synapses" on the cells that they innervate (Finger 1994). The general body plan of neurons revealed to Cajal by the Golgi stain suggested to him that the direction of impulse flow is from the dendrites, across the cell body, and then out along the axon to the terminals. Cajal reasoned that neuronal connections in the brain should consist of multiple inputs funneling into a single output, and the one distinctive fiber, the axon, looked like it should be the output. There were a few cases where knowledge of the function of the neurons made this direction of signaling obvious, such as motor neurons whose axons connect to muscle.[1] (A catalog of many of the images of neurons produced by Cajal can be found in Garcia-Lopez, Garcia-Marin, and Freire 2010; and a search of the Web will reveal a wide variety of Cajal's drawings.)

There were two alternative views of the structure of neuronal circuits under debate within neuroscience circles at the time: the neuron doctrine that held that neurons are separate entities; and the reticular theory that held that nerve fibers are connected together in a network through which nerve impulses travel unimpeded. Cajal's interpretation of his observations of Golgi-stained neurons provided strong support for the neuron doctrine. Stated in our modern nomenclature the neuron doctrine means that neurons are each enclosed in their own cell membrane and must somehow communicate with one another by mechanisms that are distinct from the mechanism of the conduction of action potentials within the neurons. The alternative, that neurons are connected together in a continuous network within which action potentials race around relatively unimpeded, creates problems for brain function. An obvious problem is that signal transmission between neurons is mostly unidirectional while action potentials can travel along an axon in either direction (see Chapter 5). How continuous networks could produce unidirectional signaling was difficult to imagine. Transmission of impulses across the synapse is a little slower than transmission along axons, which is also hard to imagine happening within a nerve net. Last but certainly not least, it is difficult to see where the choice points, where it is decided which path nerve impulses will take, would be in a continuous network, while synapses where impulses must be transferred from one neuron to the next are obvious locations for decision making.

Cajal thought that since the Golgi staining was contained within individual neurons they could not be connected together in a continuous network. This conclusion was something of a leap since the gap between neurons at synapses is so small it cannot be seen with a light microscope, and knowledge of the nature of the nerve impulse and the structure of the synapse would not be available for decades. It has turned out that the distinction between separate cellular units and cellular networks is not as absolute as Cajal thought. For example, heart muscle cells are separate cells but are attached in a network through contact points called intercalated discs where the membranes of adjoining heart cells have ion channels that line up with one another forming continuous channels through both membranes. The points of contact that form these electrical connections are called gap junctions, and the ion channels are made of proteins called connexons. The forward current loop of the action potential passes right through the gap junctions, so action potentials pass from one heart cell to the next without delay to produce the contraction during each beat (see Chapter 9). Electrical synapses containing gap junctions are formed between neurons in special cases. They were discovered in crayfish giant synapses (Furshpan and Potter 1959), and later in the mammalian nervous system where they are nowhere near as common as chemical synapses (Connors and Long 2004).

From a twenty-first-century scientific perspective, it is difficult for us to imagine how neurons could process information if they were all electrically connected within a nerve net in which the only signal is the action potential. But a lot less was known about cells in the late nineteenth and early twentieth centuries; therefore, scientists could legitimately entertain a wider range of possibilities. Not every Golgi-stained neuron that Cajal saw was entirely isolated. In some cases overlapping fibers looked like they were clearly crossing each other without any blending together of their outlines. In other cases the overlap of fibers was too great to see all of the intersections clearly enough to rule out direct connections, but Cajal said that he never saw a clear case of a fiber that could be traced to continuous branches connecting to different cell bodies. Of course, negative data doesn't mean it can't happen, and there is at least one exception—the stellate ganglion in squid (see Chapter 4). But if direct connections between neurons were common, clear examples shouldn't be that hard to find. Instead, it was easy for Cajal to find neurons that looked clearly isolated. Sometimes you just have to listen to what your data are telling you, so Cajal assumed that the cases where he could not tell definitively that fibers from different neurons were not directly connected were not counterexamples against his hypothesis— just situations that were too confused to tell what is really going on. Cajal once described himself as "an unconquerable fanatic in the religion of facts" (Finger 2000, 214). But he was also a genius in their interpretation. His reputation among neuroscientists ascended into the stratosphere where it remains to this day.

Cajal introduced several improvements in the Golgi method, but even so myelinated axons did not stain well, so he focused on neurons in developing embryos and young animals where myelin is either absent or not fully developed. Although the Golgi stain could not be applied to living tissue, Cajal was able to observe how neurons changed as the nervous system developed. The neurons he saw in embryos displayed longer and longer axons and dendrites as development proceeded, and he concluded that axons and dendrites are outgrowths of the cell body of the neuron. The growing axons were tipped with endings that spread out, looking like an amoeboid structure, which Cajal called growth cones because their shape often vaguely resembled a cone connected to the end of the axon by its apex. With a little imagination applied—and Cajal seems to have had a lot—it appeared that the growth cone crawled along, laying down the axon behind it; think of a spider spinning a web.

These apt "Cajalian" interpretations illustrate an important point about scientific progress—all scientifically drawn conclusions are provisionary and thus subject to change as more data are obtained. Nobody had ever seen axons grow when Cajal was making these observations. When the state of the art changed, confirmation was necessary. Ross Granville Harrison (1870-1959) of Yale University developed the first tissue culture system, in which live cells were maintained outside the body, specifically to observe directly how nerve fibers grow (Harrison 1910). The developing neurons were taken from animals with tissues that functioned well when removed from the body—frogs, of course. The culture medium in which frog neurons were maintained was clotted lymph, and in 1910 Harrison published his report of growth cones crawling along, laying down nerve fibers behind them, just as Cajal had envisioned. Thus, the wiring of the nervous system involves growth cones navigating through the developing embryo and forming connections with other neurons or with sensory cells or effectors such as skeletal muscle fibers.

Cajal was not one to shrink from pushing his interpretations as far as possible into even more interesting directions. He suggested that since the wiring of the nervous system changes during development by growing new connections between neurons, maybe that is also what happens during learning and memory. In 1872 Alexander Bain, a Scottish philosopher and psychologist, had originated the idea that nerve fiber and nerve terminal growth might be involved in the recording of memory (Finger 2000). The phenomenon was named neural amoeboidism. While Cajal viewed such ideas of his own and of others as highly speculative, it is now clear that anatomical changes in axon terminals at the synapse are involved in memory storage (see Chapter 8), so yet another of Cajal's insights turned out to be correct.

Camillo Golgi never accepted the neuron doctrine and carried his belief in neural networks to the grave. From Golgi's Nobel lecture, it is clear that Golgi was a detail man, requiring that the neuron doctrine explain everything he saw (Golgi 1906). During my own research career I have come to recognize a continuum of scientific temperament along several dimensions. One, perhaps the most important, is the continuum from discoverer to critic. Scientists who have ever discovered something and related the finding and its possible implications to a colleague learn quickly to brace themselves for the inevitable list of objections and apparent counterexamples that must be answered and explained. Although sometimes disappointing and even injurious to the ego, this important aspect of scientific discourse serves to reduce the proliferation of crackpot ideas, which would abound even more if it were not for some enforcement of scientific rigor. Nonetheless it can be and often is overdone.

Much of Cajal's genius arose from his feeling for how neurons should work based on what they look like. Because of his great insight, his science was truly great. Being touched by such history, even just a little, always produces a good feeling in me. While working at Cornell University, I was writing a research paper and wanted see some of Cajal's drawings in their original source. I borrowed a copy of *Histologie du système nerveux de l'homme & des vertébrés* by Santiago Ramón y Cajal, which was falling apart, from the Cornell University Library. On one of the initial blank pages at the beginning was pasted a photo of Cajal at his desk. There are many published photos of him, but this was not a professional photograph; it looked like a typical black-and-white photo taken with a Kodak point-and-shoot camera and pasted in the book, just like my family back in the 1950s pasted pictures taken with their Brownie Hawkeye in a family photo album. This was entirely possible since the Kodak Brownie camera was introduced to the public in 1900.

Figure 6.1. Snapshot of Santiago Ramón y Cajal of unknown source pasted in a copy of his book *Histologie du système nerveux de l'homme & des vertébrés* in the collection of the Cornell University Library.

I resisted the strong temptation to liberate the photograph to my private collection—after all, the page was loose—and explained to the librarian when I returned the book that it needed to be restored and placed in the rare book collection. I remembered this photo while writing my manuscript; my wife checked the Cornell University online catalog and found the book listed with a note that a photograph of Cajal is pasted inside. I hope a few others have been as thrilled to find it as I was. I asked a Cornell colleague and friend, Professor Emeritus Howard Howland, to take a picture of the photograph for me, which is shown in Figure 6.1.

THE AXON

One of the consequences of the neuron doctrine is the requirement to explain how a neuron can produce and maintain an axon. Axon terminals are complex machines that are highly dependent upon supplies produced in the cell bodies and delivered by transport through the axon to the terminals. In order to understand the mechanisms of synaptic transmission, it will be helpful to understand some of the inner workings of axons, so let's briefly dial forward to our modern level of knowledge.

When axons are growing, the growth cones need a supply of membrane lipids, membrane proteins, and internal proteins in order to construct more axon. Lipids and proteins do not last forever, and once the neurons have connected, the axons and their terminals still need a continuous supply to replace those that have been degraded. Also, we now know that the growth of axon terminals that produces new connections is an important part of the mechanism of long-term memory (see Chapter 8). Some lipids and proteins can be synthesized within the axons, but others must be synthesized in the cell bodies and transported to the sites in the axon and axon terminals where they function. For example, the major membrane phospholipids can be synthesized in the axons, but cholesterol, which is also a major constituent of cell membranes, is synthesized in the cell bodies and delivered by axonal transport (Vance et al. 1994). The major structural proteins within the axon are largely synthesized in the cell bodies, but certain proteins that have major functions within the terminals can be synthesized at their sites of function.

It was difficult to understand how anything in the cell bodies could traverse the long distances required to reach the terminals until the mechanism of axonal transport was discovered in the second half of the twentieth century. Axonal transport plays many supportive roles in animal electricity, such as supplying the axon and its terminals with ion channels and supplying the terminals with vesicles for the release of chemical neurotransmitters at synapses (see Chapter 7). The infrastructure that supports axonal transport is a system of microtubules running through the entire length of the axon, including all of its branches

(Kandel et al. 2013). Microtubules are polymers of the protein tubulin that assemble in a tight coil forming a submicroscopic tube. The hole down the middle of the tube has no mechanistic function; all of the action occurs on the surface. Microtubules form the infrastructure of one of the two great systems that produce movement in organisms. The other system utilizes a protein infrastructure of actin filaments and myosin and is responsible for the contraction of muscle cells and for a range of cell movement, including the crawling of amoebas and growth cones. The actin and myosin system in skeletal muscle is discussed in the Chapter 7. The tubulin system powers the movement of small components inside the cells called organelles. The most familiar example, which you may remember from high school biology, is the pulling of chromosomes to opposite sides of the cell by the mitotic spindle in preparation for cell division. The fibers of the mitotic spindle are microtubules.

Microtubules form tracks along which the motor proteins kinesin and dynein move. Motor proteins have as part of their structure flexible heads that swing back and forth, functioning much like oars, but instead of dipping into water, the kinesin and dynein heads bind to microtubules and then pull themselves forward with a power stroke fueled by ATP. There are many types of kinesin and dynein, which are distinguished by their direction of travel along the microtubule and by the structure of their tails. The tails are binding sites, which determine what cargo the kinesin and dynein bind to and thus carry along the microtubules. The cargo carried by kinesin and dynein is mostly tiny axonal transport vesicles with proteins in their membranes, some of which bind to the kinesin and/or dynein tails. Each transport vesicle has many kinesin and/or dynein molecules bound to its membrane, which act together to carry the vesicle along; think of the motion of the kinesin/dynein heads as the legs of the vesicle.

The microtubules are much shorter than many axons, about one mm on average, but they overlap with one another, and transport vesicles have no problem hopping from one to the next, following an uninterrupted path along the axon between the cell body and the axon terminals. Microtubules are asymmetrical from end to end; the different ends of the microtubules are designated as plus and minus (not to be confused with electric charge). Microtubules are oriented in axons with the plus ends extending in the direction away from the cell bodies. With just a few exceptions, most types of kinesin bind to microtubules in such a way that the power stroke pulls them in the direction of the plus ends of microtubules, thus propelling their cargo from the cell bodies along the axon toward the terminals in a process known as anterograde transport. Dynein moves its cargo in the opposite direction toward the minus ends of the microtubules, carrying cargo along the axons from the terminals toward the cell bodies in a process known as retrograde transport. Thus, axons contain a two-way transit system. Anterograde transport makes it possible for the cell body to supply proteins and other molecules required for the growth and maintenance of the axon. Retrograde transport carries biochemical signals within the axon from growth cones and mature axon terminals to the cell body. Among the processes that are controlled by retrograde biochemical signals are those that support the survival of axon branches and the entire neuron. These mechanisms come into play during the development of the nervous system and become especially important when neurons are injured or diseased. There will be more about this in Chapter 8.

The anterograde transport system is the source of most all the proteins that play roles in the mechanisms of animal electricity. Membrane proteins, for example, the all important sodium and potassium channel proteins, are synthesized in the cell bodies under the direction of messenger RNA that is produced in the cell nucleus in a complimentary sequence to DNA.[2] Since membrane proteins cannot dissolve in water, their synthesis occurs associated with a membrane system in the cell, the endoplasmic reticulum (ER). The hydrophobic regions of the protein are inserted into the ER membrane as the protein is being assembled. Like the cell membrane, the ER membrane encloses a lumen. The ER has a complex, flattened shape so surface area is emphasized and the lumen is a thin sliver. After membrane proteins are synthesized from messenger RNA, the tiny region of ER membrane containing them buds off, forming vesicles released into the cytoplasm. These vesicles fuse with another membrane system, the Golgi apparatus (discovered in 1898 by Camillo Golgi) where the membrane proteins undergo a further stage of processing. The output of the ER and Golgi systems enters the cytoplasm of the cell bodies in the form of vesicles budded off from the Golgi membranes. These include vesicles destined to travel in the anterograde axonal transport system. The axonal transport vesicles have kinesin on their surfaces and are transported along the axonal microtubules to sites of function where the vesicles fuse to the axonal membrane or to internal membranes in the axon. Fusion to the cell membrane involves attaching to the cell membrane and forming a pore linking the inside of the vesicle to the outside of the cell. Then the vesicles turn inside out and their membrane is incorporated into the cell membrane with what used to be the inside surface of the vesicle facing the outside of the cell. Thus, when membrane proteins such as sodium and potassium channels are synthesized in the ER, the ends of the channels that face the lumen of the ER wind up on the outside surface of the cell. When the membrane containing them turns inside out during insertion into the cell membrane the channels will be oriented in the direction required for proper functioning.

The fusion of vesicles to the cell membrane has another function besides adding membrane lipids and proteins to the surface; fusion with the surface membrane can also release the contents of the vesicles into the extracellular space. This is a pervasive mechanism for secretion of hormones into the bloodstream, digestive enzymes into the gut, and many other essential functions. One of the most important roles of vesicle fusion is the release of neurotransmitters from axon terminals at synapses. Now, let's dial back to the history of how this mechanism was discovered.

CHEMICAL SYNAPTIC TRANSMISSION

As the twentieth century progressed, the concept that transmission of nerve impulses must be electrical was gradually taking hold. Cajal had characterized neurons as individual cells surrounded by their individual cell membranes. Therefore, action potentials must be confined within them with no obvious way to signal the next cell downstream in the circuit. What happens when an action potential, after completing its trip down the axon, reaches the axon endings? Animal spirits were long gone, and the quest for the mechanism of synaptic transmission was framed as distinguishing between two alternatives—is synaptic transmission electrical or chemical? The electrical transmission camp had speed on its side. Secretion of chemicals such as hormones was known at the time and viewed as a slow process since it requires time for secretion and travel to target cells that could be a considerable distance away. As always, scientists attempt to relate the unknown to the familiar. How could a train of action potentials arriving at the nerve-to-muscle synapse at a frequency of hundreds per second produce a similar rapid train of action potentials in the muscle fibers by squirting hundreds of pulses of a chemical on it each second? The imagery just did not work. However, the chemical transmission camp had drugs on its side since a variety of chemicals had been discovered that block synaptic transmission with no effect on action potentials, suggesting to some that the transmission itself must be chemical.

The pharmacologists Otto Loewi, Henry Dale, and their colleagues carried the banner of chemical transmission to victory (Dale 1936; Loewi 1936; Finger 2000; Tansey 2006). Chemical transmission is far and away the major mechanism of synaptic transmission. However, electrical transmission between cells is also present in specialized, but very essential, circumstances such as the connections among heart muscle cells discussed above and between pacemaker cells and heart muscle that transmit a massive action potential that invades the entire heart to produce each beat (see Chapter 9). Somewhat paradoxically, the heart is also where it was first conclusively shown that synaptic transmission is through chemical transmitter substances released by the axon terminals. The axon terminals in the heart come from two sources that innervate the pacemaker cells, the cardiac accelerator nerve, the activation of which speeds the heart rate, and the vagus nerve, the activation of which slows the heart rate.

In 1904 Henry Dale began work at a drug company research institute in England, the Wellcome Physiological Research Laboratories of Burroughs Wellcome & Co. At the time professional scientific research was just beginning to develop in England, and a drug company job gave Dale unrivaled research facilities where he could pursue research unencumbered by the heavy teaching loads typical of most academic positions. But not all interactions with students are impediments to research, and Dale became aware of the discovery in 1904 by Thomas Renton Elliott, a Cambridge undergraduate, that administration of adrenaline to an animal produced effects similar to sympathetic nerve stimulation, which contracts the arteries and accelerates the heart rate. The student wrote a paper that became famous in which he speculated that the effects of nerve stimulation might be due to the release of adrenaline from the nerve endings. Dale began experiments with compounds structurally similar to adrenaline and showed that one of them, noradrenaline, produced effects that were the most similar to nerve stimulation. At the time both adrenaline and noradrenaline were just chemicals synthesized in the laboratory, and there was no indication that they were produced in organisms. It would be shown later that sympathetic axons of the cardiac accelerator nerve increase heart rate by releasing noradrenaline from their terminals onto the cardiac pacemaker cells, but that came after the demonstration that acetylcholine released from parasympathetic axons in the vagus nerve slows the heartbeat.

It is both fortunate and fitting that acetylcholine was the first chemical demonstrated to function as a neurotransmitter. It was fortunate because in 1913 Henry Dale discovered biologically produced acetylcholine by chance in extract of ergot, a fungus that infects rye. At the time ergot extract was used medically in childbirth to stimulate labor. Wellcome had assigned Dale the task of studying the pharmacological properties of ergot, which were apparently less interesting to him than his studies of the nervous system: "We got that thing out of our silly ergot extract. It is acetyl-choline and a most interesting substance. It is much more active than muscarine, though so easily hydrolysed that its action, when it is injected into the blood-stream, is remarkably evanescent, so that it can be given over & over again with exactly similar effects, like adrenaline. Here is a good candidate for the role of a hormone related to the rest of the autonomic nervous system. I am perilously near wild theorizing" (Tansey 2006, 420).

One never knows in advance where great discoveries will come from, and they sometimes emerge during the performance of undesirable tasks that, fortunately it turned out, could not be avoided. It was fitting for acetylcholine to be the first neurotransmitter discovered since it is distributed throughout the nervous system. It is the transmitter in sympathetic and parasympathetic ganglia, in parasympathetic neurons and some sympathetic neurons that innervate peripheral tissues, in motor neurons that innervate skeletal muscle, and in many neurons in the central nervous system. But that was all still to come. At the time it was unknown whether acetylcholine even occurred naturally in any animal.

World War I derailed Dale's work on acetylcholine in 1914. He turned his attention to secret medical research for the British government, likely the search for antidotes to poison gas. In 1921 Otto Loewi at the University of Graz in Austria conducted the now very famous experiment that brought Dale's attention back to acetylcholine. In the often-cited story, the definitive experiment that would provide the proof of chemical synaptic transmission came to Loewi in a dream. He woke up in the middle of the night recalling the dream and wrote the experiment down, lest he forget it by morning. But in the morning he had forgotten it and could not read his handwriting. He spent a miserable day in fear that the experiment would be lost forever. Fortunately, he awakened the next night again recalling the same dream. In the more dramatic form of the story he immediately went into his laboratory and performed the experiment, but in an alternate version he wrote it down in legible form, returned to bed, and did the experiment first thing the next morning. Either way, he obtained the dreamed of result.

His discovery could plausibly have happened this way because everything he needed was likely available in his laboratory: two frogs, the equipment to dissect out their hearts and stimulate the attached vagus nerve, and two vessels filled with a salt solution called Ringer's solution formulated to keep tissues alive and functioning. Frog hearts placed in Ringer's solution continue to beat on their own for a long while, and stimulation of the stump of the vagus nerve still attached produces a slowing of the beat rate just like vagus nerve stimulation in the intact frog. This had been known for a long time, but somehow nobody thought to do the experiment that Loewi performed. Loewi slowed the beating rate of an isolated heart for a time by stimulating the vagus nerve. He then transferred some of the Ringer's solution from the vessel that had held that heart during the stimulation to a vessel containing a second heart—this one beating without stimulation. As in the dream, the Ringer's solution taken from the stimulated heart slowed the beating of the second heart.

On their own, few experiments have one absolutely inescapable conclusion. This experiment almost certainly shows that a chemical that acts upon the heart to slow its beat must have been released into solution during vagus nerve stimulation. But further conclusions rest more heavily upon the conceptual framework from which the experiment was derived. Since Loewi's experiment was designed to detect the release of a chemical neurotransmitter from the nerve endings, it is natural to assume that the unidentified substance, which Loewi called *Vagusstoff* (German for "vagus substance"), must have been released from the terminals of the stimulated axons, but this experiment alone does not show where *Vagusstoff* came from. Maybe *Vagusstoff* was actually released by the muscle cells themselves as part of a mechanism that was completely unknown and therefore outside of the conceptual framework of choices available to Loewi. Still, Loewi had plenty of justification for celebrating that morning because his result was at the very least a major clue suggesting that *Vagusstoff* released from the axon endings carried the signal across the synapse. After all, why would something be released if not to affect something else? Upon learning of Loewi's result, Dale immediately thought that *Vagusstoff* was acetylcholine. Subsequent experiments showed that acetylcholine disappeared from the heart when the vagus nerve was cut in the frog and the axons were given time to degenerate, suggesting that the acetylcholine had been in the axons.

Synapses are sensitive to drugs, and in fact many drugs used for the treatment of psychiatric disorders and other disorders of the nervous system work through affects on synaptic transmission. Also, the evolutionary battle between the species has produced many chemical weapons such as plant poisons and snake venoms that attack synaptic transmission. The wide use of such chemicals emerged as the field of neuropharmacology. Drug effects have produced a wealth of knowledge about synaptic transmission, and it wasn't long before substantial evidence was accumulated that indicated that Loewi and Dale were both right.

The evidence came from a number of directions. First, since acetylcholine applied directly to the heart has the same effect, slowing of the beat, as vagus nerve stimulation, it was clear that a chemical, specifically acetylcholine, can produce the effects attributed to *Vagusstoff*. In our modern view this suggests that there is a receptor in the heart muscle that the chemical binds to that produces the response. More evidence came from effects of additional applied chemicals. Atropine, an alkyloid produced by plants such as deadly nightshade and mandrake, prevented both vagus nerve stimulation and acetylcholine from slowing the heartbeat. The mechanism of action of atropine was unknown at the time, but the similar inhibition of the effects of vagus nerve stimulation and acetylcholine application suggests that both are acting upon the same cause-and-effect chain of events. This turned out to be true. Atropine blocks the acetylcholine receptors in the heart. Loewi also showed that addition of acetylcholinesterase, an enzyme that destroys acetylcholine, into the Ringer's solution blocks the slowing of the heart by both acetylcholine and vagus nerve stimulation, which made it really look like *Vagusstoff* was acetylcholine. Loewi went on to show that the frog heart contains its own acetylcholinesterase, and that a cholinesterase inhibitor, eserine, prolongs the effect of vagus nerve stimulation, presumably by preventing the breakdown of acetylcholine and therefore prolonging its action. Prolongation of transmission by eserine is a powerful test for synapses that utilize acetylcholine as a neurotransmitter. It turned out to be a better test than blocking by atropine since there are two types of acetylcholine receptors, and those on skeletal muscle fibers and at many other synapses are not affected by atropine. The pharmacological investigations into the control of heart rate by vagus stimulation was an early example of the tremendously important role that drugs, most of them poisons, would play in providing clues that led to discovering the mechanisms of synaptic transmission.

In 1929 Dale and his colleague Harold Dudley identified acetylcholine occurring naturally in the mammalian body. But detection in the bathing fluid of the stimulated heart had to await the development of a sensitive assay in the form of the leech-muscle preparation by Wilhelm Feldberg, who arrived on the scene as a refugee from Nazi Germany in 1933. Like mammalian skeletal muscle, leech muscle receives its signal to contract by the release of acetylcholine from the nerve endings innervating it. This was unknown at the time, but it was known that if leech muscle without its nerve is treated with eserine, then the leech muscle becomes extremely sensitive to applied acetylcholine. With leech muscle suspended in a bath and connected to a lever that moved a kymograph stylus, the relationship between the amount of acetylcholine in the bath and the magnitude of the contraction could be accurately calibrated. These were still primitive times for scientific techniques compared to what we are used to today. Although the smoked drum kymograph was soon replaced by electronic recording methods, the leech-muscle assay remained the state-of-the-art method of detection of acetylcholine until sensitive chemical methods were developed in the

1960s. Not much was learned at this early stage about the mechanism of release of acetylcholine.

Dale took up the question of whether acetylcholine is the neurotransmitter in systems other than the heart. There was good reason to think that synaptic transmission in the heart may be different from synaptic transmission between neurons and between neurons and muscle. The heart beats independently. The slowing of the beat in response to vagus nerve stimulation is a response that occurs over a matter of seconds, not milliseconds. Therefore, the concern that secretion of a chemical may not be fast enough to be the mechanism of synaptic transmission between neurons and between neurons and skeletal muscle does not apply to the heart. It was quite possible that acetylcholine would not have worked as a neurotransmitter anywhere else. But testing showed that blocking of acetylcholinesterase with eserine prolonged the contraction of skeletal muscle in response to nerve stimulation, suggesting acetylcholine is the neurotransmitter at the motor neuron-to-skeletal muscle synapse, which we now know to be the case. Acetylcholine also functions as an excitatory neurotransmitter in some of the neurons in the brain. Among them are the neurons that are especially susceptible to degeneration in people with Alzheimer's disease.

The response of skeletal muscle to acetylcholine has to turn off almost as quickly as it turns on to allow the synapse to faithfully transmit high frequencies of action potentials arriving at the axon terminals. The short duration of the response is accomplished by the placement of acetylcholinesterase at the synapse in very close proximity to the acetylcholine receptors so that once released the acetylcholine acts for a very short time before it is degraded (see Chapter 7). Interestingly, if the acetylcholinesterase system in the heart had been as efficient, the vast majority of acetylcholine released from the endings of the axons in the vagus nerve would have been degraded, and Loewi's dream experiment would not have worked because the acetylcholine would not have accumulated in the bathing fluid of the stimulated heart. About 30 years after the experiment, Loewi pointed out that at the time he had not considered the possibility that even if transmission from the vagus nerve endings to the heart was via a chemical there might not be enough escaping into the Ringer's solution to slow the second heart. On reflection it seemed to him unlikely that enough could be released into the Ringer's solution, but he did not think of this objection at the time and went ahead with the experiment. In Otto Loewi's own words: "Sometimes in order to become a discoverer one has to be a naïve ignoramus!" (Finger 2000, 270). Loewi makes an important point; knowledge has a way of helping one talk oneself out of doing an experiment. If the possible result is interesting enough, and you don't have to sell the farm to proceed, it is sometimes better to throw caution to the wind and just do the experiment. Otto Loewi (the once-naive ignoramus) and Sir Henry Dale shared the 1936 Nobel Prize for the discovery of chemical synaptic transmission.

Some confusion ensued when others failed to reproduce Loewi's result using mammalian hearts. Mammals are different from frogs in many ways, but not in the mechanism by which vagus nerve stimulation slows the heart. It was eventually discovered that mammalian blood, which was used to bath the mammalian hearts, contains an enzyme that destroys acetylcholine; it was fortunate for Loewi that he dreamed of frog hearts. Concerns also arose when it was discovered that the vagus nerve contains acetylcholine along its length, not just at the endings. The action potential mechanism was not known at the time, and some scientists thought that acetylcholine might be involved in transmission of the action potential, which may have been responsible for its release into the Ringer's solution bathing the stimulated heart. John Eccles, the Australian neurophysiologist who would go on to share the Nobel Prize with Hodgkin and Huxley in 1963, accepted the demonstration that chemical synaptic transmission occurs, but thought incorrectly that the initial signal across the synapse was too fast and had to be electric, which was then followed up by the slower chemical component of the transmission. Confirmation that release of neurotransmitter is the principal mechanism of synaptic transmission awaited the analysis of neuromuscular transmission in the frog by Fatt and Katz in 1951 using microelectrode intracellular recording (see Chapter 7).

NEUROTRANSMITTERS GALORE

In the years since Henry Dale and Otto Loewi discovered acetylcholine, the known neurotransmitters have multiplied prodigiously, now numbering over 50. Neurotransmitters operate by binding to receptors on membranes. There are two types of neurotransmitter receptors: neurotransmitter-activated ion channels, and neurotransmitter-activated receptors that directly trigger internal biochemical mechanisms in the receiving cell. Some neurotransmitters bind to several types of receptors, some of which are ion channels and others of which trigger internal biochemical signals. The binding of acetylcholine to a receptor on the heart pacemaker cells triggers an internal biochemical mechanism that opens potassium channels, producing an outward potassium current that opposes the automatic depolarization of pacemaker cells, thereby slowing the heart (see Chapter 9). Acetylcholine released at synapses of motor axons on skeletal muscle bind to a receptor that is an ion channel. Opening of the acetylcholine receptor channel allows the small positive ions sodium and potassium to cross the muscle fiber membrane, and the entry of sodium depolarizes it to threshold, generating action potentials in the muscle fiber.

A variety of amino acids also serve as neurotransmitters. The most widely distributed excitatory neurotransmitter in the central nervous system is the amino acid glutamate. Some glutamate receptors are ion channels, which open when glutamate binds, allowing the entry of sodium that produces depolarization, while other glutamate receptors trigger internal biochemical reactions. Unlike acetylcholine, glutamate is not broken down in the synaptic gap. The glutamate signal is terminated because glutamate transporter pumps in the axon terminal and surrounding cells recover the glutamate, which is reloaded into synaptic vesicles and reused (Amara and Kuhar 1993).

Gamma-aminobutyric acid (GABA) is the major inhibitory neurotransmitter in the brain. GABA binds to the GABA$_A$ receptor, which is a chloride channel. Opening of the GABA$_A$ receptor channels moves the membrane potential closer to the chloride equilibrium potential, which is near the resting potential (see Chapter 3). The increased contribution of the chloride current to determining the membrane potential reduces the potency of excitatory synapses on the same neuron. In most neurons, the action potential is generated at the base of the axon at the cell body, called the initial segment, where voltage-gated sodium channels are present in high density and the firing threshold is consequently low. Since opening of chloride channels reduces the electrical resistance of the membrane, activation of inhibitory synapses that lie on a dendrite in the path between excitatory synapses and the initial segment of the axon will short-circuit some of the forward current generated by the excitatory synapses across the membrane, leaving less forward current and less depolarization to reach into the initial segment where it could contribute to bringing the membrane to threshold.

Acetylcholine, glutamate, GABA, and some other neurotransmitters are known as small molecule neurotransmitters to distinguish them from the neuropeptides, which are larger (see below). Small molecule neurotransmitters are the major carriers of the signals between neurons that carry the information flow in the central nervous system. These neurotransmitters link our sensory systems to the unconscious and conscious processes of our brains and ultimately transmit signals out through our muscles to generate our behavior. While the roles of these neurotransmitters in brain function overlap with one another, some specialized functions within the brain are associated with specific neurotransmitters: memory with acetylcholine, emotion with serotonin, and fine motor control with dopamine. Drugs have been developed that target specific neurotransmitters. For example, the class of antidepressant drugs known as selective serotonin uptake inhibitors work by slowing the reuptake of serotonin from the synaptic gap into the axon terminals. Like glutamate, serotonin is not broken down in the synaptic gap, so preventing its reuptake prolongs its action at the synapse, producing the antidepressant effect.

Neurotransmitters are not always released alone. A class of neurotransmitters known as neuropeptides consists of short chains of amino acids, which are released in combination with the small molecule neurotransmitters at certain synapses. There are a great many neuropeptides and their functions vary. There are also full-size proteins that transmit messages across synapses in the opposite direction of the travel of messages carried by electrical signals; that is, they carry their signals in the retrograde direction back into the axon terminal. One class of retrograde signal carriers is the neurotrophins that play roles in the establishment of neuronal circuits during the development of the nervous system (see Chapter 8). For example, the protein nerve growth factor (NGF) is released from cells into the surrounding extracellular space. The binding of NGF to receptors on growth cones can promote and direct their growth, and binding to receptors on mature axon terminals can support the survival of the neuron and the maintenance of its synaptic connections.

Another important carrier of signals across synapses, nitric oxide, is at the other end of the molecular size continuum. You may have heard of nitroglycerine being administered to people with heart pain. Nitroglycerine is metabolized by the energy-producing organelles in the cell, the mitochondria, producing nitric oxide. It is the nitric oxide that dilates the arteries in the heart. In the context of its role in signaling between neurons, nitric oxide is often described as a gas, which it is when released into the atmosphere. However, nitric oxide, oxygen, and nitrogen inside cells and in the extracellular fluid are all dissolved as individual molecules just like any other dissolved substance. Nitric oxide is produced within the receiving cell at active synapses in the hippocampus, which is a part of the brain involved in memory (see Chapter 8). Nitric oxide can cross the lipid membranes, so it cannot be and is not stored for release in vesicles. Instead, immediately after it is produced, nitric oxide diffuses in the retrograde direction across the synaptic gap and into the axon terminals at active synapses where it stimulates the growth of tiny new axon branches that give rise to additional terminals, thereby increasing the neurotransmitter that can be secreted with each arriving action potential. This strengthens the synapse and is one of the mechanisms of learning (see Chapter 8).

You can see that the field of neurotransmitters has reached enormous complexity. The knowledge that has been gained of transmission in all the different kinds of synapses is nowhere near the detailed and experimentally verified status of transmission at the motor neuron–to–skeletal muscle synapse described in Chapter 7. I am sure most neuroscientists believe that most of the major principles of central synaptic transmission are understood, but it is the central nervous system after all, and there is undoubtedly plenty of room for tremendously important variations, some of which we may not even imagine. Paraphrasing the words of one of my postdoctoral supervisors, David Potter: We are ignorant. Never forget that although we know a lot about the nervous system, there is much more that remains unknown.

7

NERVE TO MUSCLE

When you decide to move your left little finger to press down and type the letter *a* on your computer, the pattern of action potentials traveling down the axons in your arm constitutes the complete message that tells each muscle fiber involved in that act when to contract, how much, and for how long. The muscle fibers that produce this movement are just following orders. They do not participate in the decision. If we could simultaneously detect and knew how to read the pattern of action potentials in all your motor axons, we could reconstruct your movements precisely just from that information. This has been done, for example, in lamprey eels. Their brain and spinal cords and peripheral nerves can be kept alive after removal from the body. When the spinal cord is electrically stimulated in a way that produces swimming in the intact animal, the same pattern of action potentials is produced in the motor axons that exit the spinal cord, indicating that the lamprey's nervous system is taking a swim even though there are no muscles to do the swimming (Wallén and Williams 1984). The function of transmission from nerve to muscle is to deliver an amplified signal that produces small contractions of the muscle fibers in response to the arrival of each and every action potential. When action potentials arrive at high frequency, the small contractions blend together to produce a smooth, powerful contraction of the muscle.

The place where the mechanism of synaptic transmission was waiting to be revealed was the nerve-to-muscle synapse, called the neuromuscular junction. It is the most accessible of synapses; Jan Swammerdam and Luigi Galvani were removing live muscle with the nerve attached from frogs centuries ago. When glass microelectrodes that could penetrate the cell membrane were developed that made it feasible to measure the membrane potential, the stage was set to figure out how synaptic transmission works.

During the same trip to Chicago during which Alan Hodgkin saw Kacy Cole's voltage-clamp data (see Chapter 4), he learned of another revolutionary technique while visiting the laboratory of Ralph W. Gerard. Somehow there had to be a way to make intracellular recordings in neurons and muscle fibers of ordinary size, that is, not just in squid axons. Gerard and his student G. Ling had found a way that established the state of the art for the next 30 years. The technique of intracellular microelectrode recording involves impaling an intact cell with a fine glass tube filled with an ionic solution (Figure 7.1). The fine tube is made by heating a larger tube and pulling it apart lengthwise to draw out two electrodes with fine tips, each a fraction of a micron in diameter, but still with small openings. Ling and Gerard pulled the fine glass pipettes by hand, which required much finesse, but eventually automatic pipette pullers were developed that applied just the right tension and just the right heat from a coil surrounding the tube to pull electrodes with just the right sized tips. The need for finesse then shifted to adjusting the parameters of the pipette puller, and many investigators were undoubtedly driven close to insanity trying to get electrodes that would work just right. Finesse was also required to advance the micropipette mounted on a mechanical micromanipulator through the cell

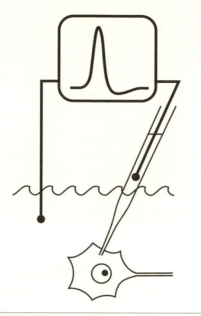

Figure 7.1. An intracellular microelectrode inserted into the cell body of a neuron.

membrane. As luck would have it the membrane seals around the electrode tip, so neither the cell contents nor too much electric current can leak out. With a few exceptions such as a giant synapse in crayfish, axon endings are generally too small to impale with a microelectrode, and most of the definitive analytical work was done on synapses in vertebrates where nerve connects to muscle. Intracellular recordings made from muscle fibers produced an amazing amount of detail about how all synapses work. Again, frogs were major contributors to human knowledge.

Hodgkin learned how to perform this technique from Ling and Gerard, and when he went back to Cambridge he conducted experiments in collaboration with an American visitor, William L. Nastuk, that showed that frog muscle fibers have overshooting action potentials driven by the sodium equilibrium potential, just like squid axons. This was a valuable contribution because of the concern that studies of squid axons may not yield data that are relevant to vertebrate nerve and muscle. The paper was scheduled to appear in the *Journal of Cellular and Comparative Physiology*, possibly before Ling and Gerard's paper appeared in the same journal. Appropriately, Hodgkin and Nastuk asked the journal to hold off publication of their own paper until after Ling and Gerard's paper was published, and the journal complied. To have not done so could have created uncertainty in the future about who invented this revolutionary technique. Glass intracellular microelectrodes became the state of the art for studying electrical activity in cells for three decades. Cell size was still an advantage in making intracellular recordings with glass microelectrodes; mammalian axons were still outside of the range, but muscle fibers, which are much larger in diameter, were perfect.

Scientific discussions of synaptic transmission are loaded with terms and acronyms. It is impossible to avoid using a few of them. The synapse between an axon and a muscle fiber gets a special name, the "neuromuscular junction." The muscle membrane under the axon terminals at the neuromuscular junction is called the "motor endplate" or just "endplate," and the change in membrane potential of the muscle fiber in response to an action potential arriving at the axon terminals is the "endplate potential." Using these terms can lead to confusion, since they apply only to synapses upon muscle fibers, and there is a list of other terms that refer to all synapses. The latter list is more useful to us, so wherever possible we will use the more general terms. Everything on the nerve terminal side of a synapse is referred to as "presynaptic," and everything on the receiving side of the synapse is referred to as "postsynaptic." The nerve terminal membrane from which neurotransmitter is released is the "presynaptic membrane," and membrane containing the neurotransmitter receptors is the "postsynaptic membrane." The presynaptic and postsynaptic membranes are separated by the "synaptic gap," across which the released neurotransmitter must diffuse to reach the receptors. In synapses where the receptors are neurotransmitter-activated ion channels such as the neuromuscular junction, the response of the postsynaptic membrane to neurotransmitter is called a "postsynaptic potential." In muscle fibers, the "excitatory postsynaptic potential" (EPSP) depolarizes the postsynaptic membrane to threshold, generating action potentials in the muscle fiber.

The neuromuscular system is an amplifier, and amplification of the signals coming out of the spinal cord begins in the axons of the motor neurons. In the motor systems of vertebrates, each muscle fiber is innervated by a single neuron, and when that neuron fires an action potential, all of the muscle fibers that it innervates are stimulated to contract. To understand the amplification that takes place, consider a mammalian example. The soleus muscle in the leg of a cat contains about 25,000 muscle fibers innervated by axons from about 100 motor neurons (Nicholls et al. 2001), whose cell bodies reside in the spinal cord. Multiple axon branches arise from the axon of each motor neuron by successive bifurcation. Since each muscle fiber is innervated by just one motor neuron, there must be an average of 250 axon branches from each neuron to innervate all 25,000 muscle fibers, but there is wide variability with some motor neurons innervating many fewer than 250 muscle fibers and others innervating many more. When the brain signals the muscle to produce a contraction of maximum force, high-frequency trains of action potentials travel down the 100 axons from the spinal cord. Once in the vicinity of the muscle fibers, the action potentials in each axon turn into two action potentials at each bifurcating branch point. This happens over and over again so that 100 action potentials, one in each axon, turns into an army of 25,000 action potentials arriving nearly simultaneously—one action potential approaching the axon terminals at each muscle fiber. Axon terminals on a single muscle fiber consist of several more axon branches, all connected to terminals close together on the muscle fiber, so the figure of 25,000 action potentials actually doubles a few more times before reaching the terminals and causing the release of acetylcholine. It is likely that upward of 100,000 action potentials arrive at the neuromuscular junctions after a single firing of the 100 neurons innervating the soleus muscle. This would produce a brief twitch. A useful contraction of the muscle requires a sustained force produced by trains of action potentials arriving at the neuromuscular junctions at frequencies as high as several hundred per second, so the action potentials arriving during normal activity easily reaches millions per second. It is hard to imagine how many billions of action potentials were required to make my hands type this paragraph.

Another stage of amplification occurs with transmission across each neuromuscular junction. The work of British neuroscientists Paul Fatt and Bernard Katz launched the discovery of the detailed mechanism of synaptic transmission. In their words: "A very large amplification of ionic currents . . . must occur at the point where an impulse is transferred from minute nerve endings to the enormously expanded surface of the muscle fibre" (Fatt and Katz 1951, 352). Even if there were direct electrical connections between the axon terminals and the muscle fiber, it is unlikely that the inward current generated by the open sodium channels in axon terminals could generate enough forward current to depolarize the muscle fiber membrane to threshold; the current generated by the nerve endings would be spread too thinly throughout the comparatively vast volume of the muscle fiber to depolarize the membrane potential to threshold at any point. The argument that transmission between neurons and between neurons and muscle fibers had to be electrical because it was too fast to be chemical (see Chapter 6) did not give this issue its due importance. Nor did it take into account that events on the nanoscopic and molecular scales can actually take place far faster than we creatures who experience the world on a larger scale can easily imagine.

FATT, KATZ, AND CURARE

Peering into an unseen world for the first time can produce a breathtaking leap of understanding. The leap is sometimes premeditated as in Millikan's discovery of electrons sitting in plain "view" on oil droplets (see Chapter 2), but perhaps more often than not the window to an unseen world appears as a surprise right in front of the experimenter's nose. The window into the nanoscopic scale of synaptic transmission was opened by the development of the intracellular microelectrode. When Fatt and Katz observed transmission across the neuromuscular junction with microelectrodes inserted in frog muscle they immediately got a view of vesicles of neurotransmitter fusing with the presynaptic membrane and releasing neurotransmitter, but it was some time before the information would be available that would tell them what they were actually looking at.

Fatt and Katz used microelectrodes to reveal the fundamental process of neurotransmitter release in our old friend the frog. From the start they knew that acetylcholine was released from the axon endings, then bound to receptors on the muscle fiber, and ultimately degraded by acetylcholinesterase (see Chapter 6). Previous work with extracellular electrodes had revealed about a 1-millisecond synaptic delay between the arrival of an action potential and the electrical response of the muscle fibers. Synaptic delay is the extra time it takes the signal to cross the synapse above the time taken by the action potential to travel the same distance along the axon.

Upon inserting a microelectrode into a frog muscle fiber near the neuromuscular junction, Fatt and Katz first immediately observed a drop in voltage to a resting potential of about −90 mV. Stimulating the axon to produce an action potential resulted in a complex depolarization of the muscle fiber, which looked like more was going on than just an action potential. The depolarization was produced by the combination of the EPSP, which rises and lasts longer than the action potential, with the action potential riding on top.

Treasures are where you find them, which is why the diversity of life on earth has been so important to science and medicine. Much knowledge and many medical treatments have been derived from poison. Hundreds of years before Luigi Galvani was trying to determine whether animals are electric machines, the Macusi Indians of Guyana were hunting with blowgun darts whose tips had been coated with a poisonous extract of a plant, *Strychnos toxifera*. In the early nineteenth century, explorer Alexander von Humbolt discovered how the Macusi prepared the poison, now known as curare. A description of the procedure was included in his massive multivolume work on Latin America (Hoffman 2009).

Curare blocks the contraction of all skeletal muscles, some of which we, and every other air-breathing vertebrate, use every time we take a breath. It does not interfere with the heartbeat, and experiments conducted in the early 1800s showed that animals poisoned with curare could be kept alive by artificial respiration and completely recovered after the effects of the drug had worn off.

THE EXCITATORY POSTSYNAPTIC POTENTIAL

Curare blocks muscle contraction by competing with acetylcholine for binding to the acetylcholine receptor channels. When curare binds, the channels remain closed in the presence of acetylcholine. Fatt and Katz applied enough curare to block some, but not all, of the acetylcholine receptors, which reduced the amplitude of the EPSPs to a level below the

Excitatory postsynaptic potentials

a at the synapse

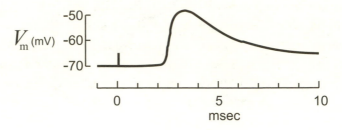

b 1 mm from the synapse

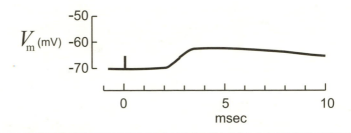

Figure 7.2. Depiction of excitatory postsynaptic potentials in a muscle fiber recorded at the synapse and one millimeter away from the synapse.

threshold for generating action potentials. Thus, when the nerve was stimulated, no action potential was produced, and the muscle did not contract, so the amplitude of the EPSP, albeit reduced by the drug, could be measured and its shape revealed. When recorded near the postsynaptic membrane, the EPSP looks like a broad, low wave a bit longer in duration than the action potential (Figure 7.2a). The long tail of the EPSP reflects the passive decay of the current across the membrane capacitance that lasts for a time after the acetylcholine receptor channels have closed. Electrodes inserted at progressively greater distances from the postsynaptic membrane recorded EPSPs progressively lower in amplitude and broader in duration (b).

It was immediately clear to Fatt and Katz that the EPSP was produced by inward ionic current across the postsynaptic membrane. After

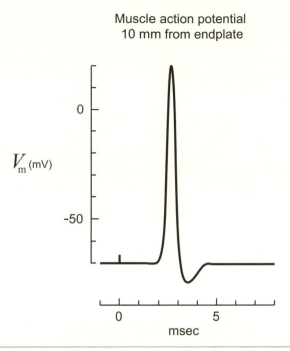

Muscle action potential
10 mm from endplate

V_m (mV)

0

-50

0

5

msec

Figure 7.3. Depiction of an action potential recorded with an intracellular microelectrode in a muscle fiber at a site away from the influence of the synaptic potential.

entering the muscle fiber the current spreads along the muscle fiber in both directions. The depolarization becomes progressively less with distance as the current dissipates across the membrane capacitance. The voltage-gated sodium channels in the muscle membrane that are responsible for propagating the action potential are distributed throughout the muscle membrane, including the area immediately surrounding the postsynaptic membrane. Under normal conditions the rise of the EPSP in the surrounding region quickly exceeds the threshold and triggers an action potential (Figure 7.3). Neuromuscular junctions are ordinarily located at about the midpoint of the muscle fibers, so actually two action potentials are generated: one traveling away from the neuromuscular junction in each direction. The localization of the EPSP to the vicinity of the neuromuscular junction convinced Fatt and Katz that the opening of acetylcholine receptor channels is controlled differently from the opening of the voltage-gated sodium channels that produce the action potential. The open versus closed state of the acetylcholine receptor channels is unaffected by membrane voltage. The acetylcholine molecules and a few other molecules, including a range of drugs, control the opening and closing of the acetylcholine receptor channels.

Mathematical analysis of the spread of EPSPs away from the postsynaptic membrane was possible because the muscle fibers are cylinders and the spread of current and voltage is determined by the internal resistance of the muscle fiber and the resistance and capacitance of the membrane, all of which could be measured. Fatt and Katz found that the spread of the EPSP fit the predictions of the mathematical analysis. Also, by inserting two microelectrodes into a single muscle fiber, Fatt and Katz showed that a pulse of inward current delivered through one electrode produced a spread of voltage similar to the spread of voltage around the postsynaptic membrane during the EPSP. Thus, it was concluded that the acetylcholine released during neuromuscular transmission opened enough acetylcholine receptor channels in the postsynaptic membrane to depolarize the surrounding muscle membrane to threshold, thus triggering action potentials that move in both directions along the muscle fiber and trigger contraction.

Fatt and Katz calculated that 2.4×10^{-14} moles of charge are transferred across the postsynaptic membrane and surrounding muscle membrane during a single normal EPSP. This is nearly the amount of charge transferred by sodium across one square millimeter of axon membrane during an action potential—way beyond the current generating capability of the axon terminals at the neuromuscular junction. This calculation supported the view that nerve endings just do not have the power to generate enough current to directly depolarize a muscle fiber. So the amplification provided by chemical synapses is essential. Somehow the release of acetylcholine at the neuromuscular junction serves to amplify the signal produced by the nerve terminals from a single axon to a level where it generates an action potential in the much larger muscle fiber. The amplification occurs because acetylcholine binds to and opens ion channels in the muscle fiber membrane, and the energy stored in the ionic concentration gradient across the muscle fiber membrane provides the power to depolarize the membrane to threshold and generate an action potential. Thus, the nerve endings only need to provide the trigger for generating the EPSP.

Fatt and Katz hypothesized that the EPSP is generated by a general increase in the ionic permeability of the postsynaptic membrane, as had been suggested by Julius Bernstein for the action potential back in 1902 (see Chapter 4). They were closer to being right since the increase in ionic permeability that produces the depolarization during the EPSP is not to a specific ion, as in the case of sodium for the rise of the action potential, but any small ions with a positive charge; that is, sodium and potassium, but not calcium, which is too large. They also noted that the acetylcholine receptor channel opening was unresponsive to changes in membrane potential; the acetylcholine receptor channels have a gate upon which membrane voltage has no effect and is instead controlled by binding of acetylcholine (and a few other molecules). The acetylcholine receptor channel is in a different class from voltage-gated channels—ion channels that are opened by chemical binding.

In its bare bones the concept of chemical synaptic transmission is simple enough: neurotransmitter is released from the axon terminals, diffuses across the synaptic gap, and binds to receptors on the postsynaptic cell, be it another neuron or a muscle fiber, causing ion channels to open. But a little thought reveals that it cannot be that simple. At the vertebrate neuromuscular junction, each action potential arriving at the axon terminal produces an action potential in the muscle fiber it innervates at frequencies that can reach upward of 400 per second. This means that the entire mechanism of release of neurotransmitter, diffusion across the gap, binding to receptors, and production of the muscle action potential must take place and the mechanism must be reset and ready to do it all over again in a few milliseconds. How this can happen 400 times in one second boggles the mind. The key is to think on the nanoscopic scale, where things we normally think of as happening very slowly happen very fast.

One of the exciting things about doing experiments is the possibility of finding something that could never have been predicted. The experiments with the frog neuromuscular junction quickly led Fatt and Katz into the nanoscale. During intracellular recordings from frog muscle they always observed a background of brief depolarizations of the membrane up to about 1 millivolt (mV) in amplitude that appeared to occur spontaneously, and more or less randomly, without nerve stimulation (Figure 7.4). These tiny depolarizations looked like miniature versions of the EPSPs produced by nerve stimulation (Fatt and Katz 1951). They seemed to arise from the release of acetylcholine for several reasons:

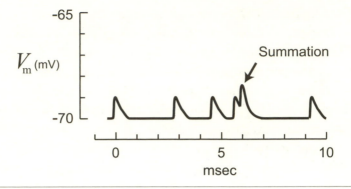

Miniature endplate potentials

Figure 7.4. Depiction of spontaneous miniature endplate potentials recorded in muscle.

miniature EPSPs were only detected by electrodes placed within a millimeter or so of the axon terminals; they were never observed when the nerve had been precut while still in the frog, and the axon endings had been allowed to degenerate before removing the nerve-muscle for the experiment; and application of prostigmine, a drug that greatly amplifies and prolongs the EPSP by blocking acetylcholinesterase, also amplified and prolonged the miniature EPSPs. Interestingly, when two miniature EPSPs happened to overlap in time, their amplitudes summed together (Figure 7.4), and when miniature EPSPs were prolonged by application of prostigmine, the depolarization produced by the summation of several EPSPs occasionally reached threshold, producing an action potential. Thus, miniature EPSPs behaved very much like they were caused by acetylcholine spontaneously released by the axon terminals interacting in the normal way with the acetylcholine receptor channels in the postsynaptic membrane. Since miniature EPSPs were recorded when the electrode placement was a millimeter away from the postsynaptic membrane, they could not have resulted from damage to the axon endings caused by the electrode. Then what was causing the spontaneous release of small puffs of acetylcholine into the synaptic gap?

Fatt and Katz found that reducing the external calcium concentration reduced the amplitude of the EPSP. Further experiments led Fatt and Katz to conclude that an action potential arriving at the axon endings causes calcium ions to enter the terminals through voltage-gated calcium channels that are opened by depolarization of the terminal. The calcium ion concentration inside the cytoplasm of all cells, neurons included, is maintained at a very low level (10–100 nanomolar in vertebrate neurons: a nanomole is 1 millionth of a millimole), and the calcium in the extracellular fluid is much higher (2–5 mM in vertebrate extracellular fluid), so the calcium concentration gradient is in the same direction as the sodium concentration gradient. While the influx of calcium contributes to the depolarization of the axon terminals, the real function of opening voltage-gated calcium channels is to allow enough calcium to enter to substantially raise the calcium concentration in the axon terminals, which triggers the release of acetylcholine.

Fatt and Katz now had means to reduce the amplitude of the EPSP in two ways. They could either reduce the amount of acetylcholine released by reducing the external calcium concentration, or they could keep the amount of acetylcholine released the same while reducing the response of the acetylcholine receptors by applying curare. As discussed above, curare reduces the effectiveness of the released acetylcholine by competitively binding with the acetylcholine receptor channels, so the same amount of acetylcholine opens fewer channels.

Application of curare at a concentration that partially blocks the acetylcholine receptor channels looks like turning down the gain of the system; a train of EPSPs displays a uniformly reduced amplitude, and the amplitude of the miniature EPSPs is also reduced by exactly the same fraction. This is as it should be if the reduction in amplitude of both EPSPs and miniature EPSPs is produced by blocking a fraction of the acetylcholine receptor channels.

But the reduction of amplitude of EPSPs in a muscle fiber bathed in low calcium is accompanied by a striking increase in the variability of the amplitudes from one EPSP to the next but the amplitude of the miniature EPSPs was not affected at all. When Fatt and Katz reduced the calcium almost to the level where EPSPs were completely blocked, some of the action potentials arriving at the axon terminals produced no EPSP at all. Some produced EPSPs of similar amplitude to the miniature EPSPs, some produced EPSPs about twice the miniature EPSP amplitude, and some three times. Very few action potentials resulted in EPSPs greater than three times the average miniature EPSP amplitude.

You may have noticed a striking similarity of this pattern of results to the results of the Millikan oil-drop experiment (Chapter 2). Instead of counting the number of electrons sitting on oil droplets, the Fatt and

Katz neuromuscular junction experiment seemed to be counting equivalent unit doses of acetylcholine released from the axon terminal. It looked as if a single unit dose produced a miniature EPSP, and the EPSP produced by the arrival of an action potential was made up of the sum of the depolarizations produced by the near-simultaneous release of a large number of doses. Fatt and Katz estimated that the increased calcium in the nerve terminal normally produced by the arrival of an action potential caused the near-simultaneous release of about a hundred equivalent doses of acetylcholine. Borrowing some of the rigor of physics for biology, they called these doses quanta.

How are quanta produced? Application of known quantities of acetylcholine to the muscle showed that the dose required for a measurable depolarization is much more than a single molecule, so a quantum is not a single acetylcholine molecule. Since the innervation of a single muscle fiber consists of fine branches leading to many fine terminals that connect at the postsynaptic membrane, Fatt and Katz considered the possibility that a quantum is the acetylcholine released by a single terminal, but there didn't seem to be a sufficient number of terminals connected to a single muscle fiber to produce the simultaneous release of anywhere near one hundred quanta. Their 1951 paper leaves the reader with the idea that each terminal may have a number of sites where acetylcholine is released and that a miniature EPSP is the spontaneous release from a single release site. But this hypothesis seems purely speculative, like the strings and valves that Descartes thought responsible for the release of animal spirits from the brain. The difference was that Fatt and Katz, being modern scientists, knew that hypotheses needed to be supported with evidence before they are believed. But what they really needed to count was not discovered until four years later.

SYNAPTIC VESICLES

The genius of the oil-drop experiment by Millikan (see Chapter 2) was in making visible the impact of a single electron; the visible signature of a single electron was the flotation by an electric field of a tiny oil droplet observable with a microscope. Without realizing it at first, Fatt and Katz had discovered the electrical signature of a single synaptic vesicle fusing with the presynaptic membrane and releasing the single dose of acetylcholine it contained into the synaptic gap. Fatt and Katz had brought modern electrical analysis to the frog nerve-muscle system that had been originated by Jan Swammerdam and utilized by Luigi Galvani for his discovery of animal electricity. The next steps were to bring the powers of electron microscopy and biochemistry to peer into the nanoscale.

In 1955 Eduardo De Robertis and Stanley Bennett observed with the electron microscope vesicles just beneath the presynaptic membrane of axon terminals in frog sympathetic ganglia and in ganglia in earthworms (De Robertis and Bennett 1955). Accordingly, they named them synaptic vesicles. Just as at the neuromuscular junction, the neurotransmitter in sympathetic ganglia is acetylcholine. The existence of synaptic vesicles, poised as if ready to release neurotransmitter into the synaptic gap, suggested that a quantum of acetylcholine is the amount contained in a single vesicle, and that synaptic transmission involves vesicles dumping their contents into the synaptic gap.

Biochemistry is in many ways the antithesis of physiology; physiology seeks to watch how things work, and biochemistry seeks to grind things up and figure out how they work by analyzing what they are made of. Combined together and with the support of anatomy, they are a formidable power for discovery. Biochemical analysis frequently requires a lot of material to work with, and the axon terminals containing the synaptic vesicles are a tiny fraction of the volume of a muscle. Not surprisingly, brains are much better sources of synaptic vesicles.

Grinding up rat brains to break up the cells and examining the wreckage with the electron microscope revealed objects that looked like broken-off axon endings. It looked as if the membranes had sealed after the axon endings broke off, and the synaptic vesicles were retained inside. A piece of membrane was observed adhering to the outside of many of the broken-off terminals. It is likely that this was a fragment of postsynaptic membrane that had adhered to the axon terminal when the cells were broken up, since structural proteins that extend across the gap hold synapses together. Structures such as axon terminals released by cell grinding can be suspended in a high-density liquid and then purified by centrifugation at high speed. Purified axon terminals can be broken down further and centrifuged to produce purified synaptic vesicles. Fractions containing purified synaptic vesicles were found to contain high concentrations of acetylcholine and choline acetyltransferase, the enzyme that synthesizes acetylcholine from choline. The fractions of purified broken-off axon endings contained acetylcholinesterase, which we now know to be bound within the synaptic gap (De Robertis et al. 1963). The brain contains many neurotransmitters in addition to acetylcholine, so most of the broken-off axon endings and synaptic vesicles purified by this procedure likely did not contain acetylcholine, choline acetyltransferase, or acetylcholinesterase, but their presence in the purified fraction indicates that a subpopulation did. Thus, it looked like some of the synaptic vesicles contained acetylcholine and that the dose of acetylcholine inside a single vesicle is, indeed, a quantum.

As with many of the processes discussed in this book, fusion of synaptic vesicles with the axon terminal membrane is a matter of probability. Arrival of an action potential increases the calcium concentration inside the terminal, which in turn increases the probability of release, causing many vesicles to fuse with the membrane and release their dose of acetylcholine nearly simultaneously. When there are no action potentials arriving, the probability of release is not zero, so in the absence of an action potential there is occasional fusion of a vesicle with the membrane producing a spontaneous miniature EPSP.

The vertebrate neuromuscular junction is the machine gun of synapses. Using ballpark figures, an action potential arriving at the axon terminal on a single muscle fiber releases about 250 synaptic vesicles of acetylcholine. At a firing rate of 300 action potentials per second, the number of vesicles releasing acetylcholine climbs to 75,000 vesicles per second—a veritable mountain of cartridge cases in a single axon ending!

To get an idea of the disposal problem, synaptic vesicles at the vertebrate neuromuscular junction measure about 40 nanometers (nm) across, so the surface area of a single vesicle comes to about 0.005 square micrometers (μm^2). Therefore, for EPSPs composed of 250 quanta of neurotransmitter, 250 vesicles fuse with the plasma membrane potentially adding 1.25 μm^2 to the surface of the postsynaptic membrane for each action potential arriving at the axon terminal. This is a large amount to add to a tiny axon ending. Since EPSPs can arrive at frequencies in the hundreds per second the addition of membrane can get very large very fast. The fusion of 75,000 vesicles during one second of firing at 300/sec would add 375 μm^2 to the presynaptic membrane of a single axon terminal.

For a neuron innervating 100 muscle fibers, one second of firing at 300/sec would add 37,500 μm^2 of membrane to the surface of the neuron. During sustained activity such as walking, motor neurons frequently fire for a lot longer than 1 second, and 60 seconds of firing would add a whopping 2.25 mm^2 of membrane to its surface! Clearly, if the membrane of the vesicles were to be somehow destroyed when neurotransmitter is released, replacing it would put an impossible metabolic burden on the neuron (Bittner and Kennedy 1970). The obvious conclusion is that synaptic vesicles that have emptied their contents into the synaptic gap must be recycled and reloaded with acetylcholine to release again. The machine-gun synapse cannot throw away the cartridge cases; they must be reloaded, and fast!

The mechanisms of synaptic transmission were revealed over a period of time during which new techniques of imaging were developed and applied to the problem. Transmission electron microscopy played a central role because synaptic vesicles are beyond the resolution of light microscopy. One limitation is that electron microscopy can only be used on dead tissue and dead cells. In the case of transmission electron microscopy the tissue is sliced into thin sections, which are treated with a heavy metal stain. Variations in the density of the metal produced by differential staining of the cell membrane and structures inside the cell block the passage of an electron beam through the tissue to varying degrees, resulting in a picture of cellular structure that cannot be seen through a light microscope. In scanning electron microscopy, tissues are not sliced into sections. Rather the specimen is coated with heavy metal and then scanned with an electron beam, forming a three-dimensional picture of the surface. A familiar use of scanning electron microscopy is the production of fantastic three-dimensional pictures of insects and other small organisms.

A special technique for preparing cells for scanning electron microscopy called freeze fracture involves freezing the tissue, physically breaking it apart, and coating the surface with metal, which is then examined by scanning the broken surface with an electron beam. If you want to demonstrate the principle for yourself, cut a thin slice of butter and prop it across one of the chambers in an ice cube tray and then add water so that when frozen the ice cube contains a butter partition across the middle. Take out the ice cube with the butter partition and strike it lightly with a hammer. If you have hit it hard enough but not too hard, it will break along the butter partition leaving halves of mostly intact ice. The ice stays largely intact because the charges on the polar water molecules (see Chapter 3) hold the ice together, while the fat that makes up butter comes apart much more easily even when cold. Likewise, tissues frozen in liquid nitrogen and then impact-fractured break apart along the lipid cell membranes rather than through the ice-filled interior of the cells and extracellular space. The lipids in the cell membrane are not exactly like butter. Since the heads of the phospholipids that make up the cell membrane are polar, the phospholipids stay attached to the ice inside and outside of the cell and the membrane separates down the middle of the lipid bilayer. Thus, the submicroscopic contours of the surface membranes are displayed in the metal mold made from the surfaces of the freeze-fractured tissue. (I'm not making this up, and it gets better!)

John Heuser, Thomas Reese, and their colleagues used freeze fracture to observe synaptic vesicles fused to the inside surfaces of axon terminals during synaptic transmission (Heuser et al. 1979). Since all electron microscopy is done on dead tissue, some ingenuity had to be applied. The greatness of the frog nerve-muscle system came through again. The frog neuromuscular junction is well suited to electron microscopy because the axon terminals are fairly long cylinders that run parallel to the muscle fiber for a considerable length. This makes the terminal membranes relatively easy to find in freeze-fractured specimens. Heuser and Reese found a way to visualize the fusion of vesicles in an especially convincing manner. They constructed a machine that instantly froze the frog muscle immediately after electrically stimulating the nerve in order to catch synaptic vesicles in the act of fusion with the presynaptic membrane.

To accomplish this they designed and built an apparatus in which the frog nerve-muscle, with stimulating electrodes located on the nerve, was attached to a platform connected by a sliding mechanism to vertical rails. The platform faced down and the frog muscle was stretched over its underside. When the platform dropped, the frog muscle came into contact with a copper block connected to a tank of liquid helium that cooled the block to 4 K (−269°C). The mechanical arrangement provided for stopping the fall of the platform supporting the frog muscle with gentle pressure against the block so the frog muscle was instantly frozen without being crushed. A magnetic switch placed along the rails was actuated by a magnet attached to the falling platform, turning on the nerve stimulation a few milliseconds before the muscle was frozen. Thus, the axon terminals were frozen in the act of releasing acetylcholine—very cool! Heuser and Reese and their colleagues then freeze-fractured the muscle and looked for impressions of the presynaptic membranes of axon terminals on the surface of the muscle fibers. Because of the convenient anatomy of the neuromuscular junction of the frog, presynaptic membranes were easily found. The top panel in Figure 7.5 shows the presynaptic membrane of a frog neuromuscular junction that was stimulated just 3 milliseconds (msec) before freezing, which is enough time for the action potentials to have reached the presynaptic terminals, but not quite enough time for the release of acetylcholine. The two lines of particles are membrane proteins likely to be calcium channels located along the presynaptic membrane. The bottom panel in Figure 7.5 shows the presynaptic membrane of another neuromuscular junction. This one was stimulated 5 msec before freezing, which is enough time for the release of acetylcholine. The pits along the presynaptic membrane such as the one boxed look like the openings that would be created by synaptic vesicles fusing with the presynaptic membrane. The arrows point to shallow dimples in the presynaptic membrane that likely represent fused vesicles at later stages when their incorporation into the presynaptic membrane is nearly complete. This is the closest anyone has ever come to seeing the fusion of a synaptic vesicle to the membrane of a nerve terminal.

It is now known that membrane proteins in the synaptic vesicle membrane and the presynaptic membrane are involved in attaching a synaptic vesicle to the presynaptic membrane and causing the membranes to fuse and release neurotransmitter. One of these, called synaptotagmin, is the link to calcium. In the absence of calcium, synaptotagmin blocks release of neurotransmitter, but calcium binding to synaptotagmin releases the block and vesicle fusion and release of neurotransmitter proceed (Nicholls et al. 2001). Calcium that enters the axon terminal is rapidly taken up by mitochondria. When the axon terminal repolarizes after the action potential, terminating the entry of calcium, the mitochondria quickly bring calcium in the cytoplasm back to

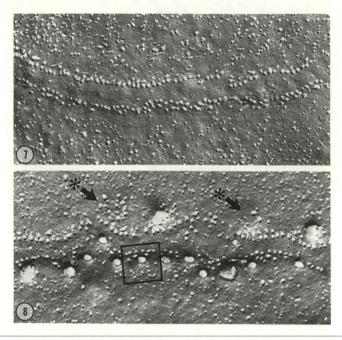

Figure 7.5. Freeze-fracture images of synaptic vesicles fused to the presynaptic membrane. (Reproduced with permission: Figures 7 and 8 from Heuser et al. 1979; © 1979 Rockefeller University Press. *Journal of Cell Biology* 81:275–300.)

its original low level, and vesicle fusion and neurotransmitter release stop, except for the occasional miniature EPSP.

Heuser and Reese (1973) explored the fate of the synaptic vesicles that had fused to the presynaptic membrane and released neurotransmitter. They used a tracer enzyme added to the solution bathing the frog nerve-muscle to track the recycling of synaptic vesicles. The tracer known as HRP (horseradish peroxidase) is an enzyme derived from the roots of the horseradish plant. HRP utilizes peroxide to oxidize a variety of substrate molecules. When HRP is taken up by a tissue, its location inside of cells can be determined by supplying peroxide and a substrate such as diaminobenzidine; the oxidation of diaminobenzidine results in a dark brown product, which can be visualized using electron microscopy. HRP is taken up into the axon terminals of electrically active neurons. So after the living frog nerve-muscle was bathed in a solution containing HRP for a set time during which the nerve was electrically stimulated, the HRP substrate was added to produce stainable prod-

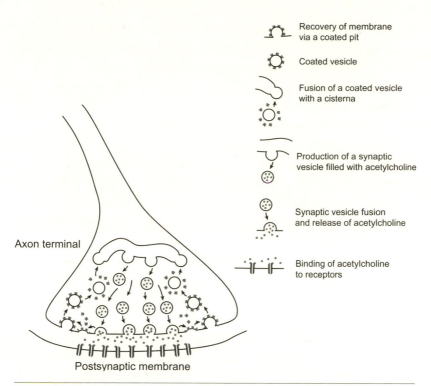

Recovery of membrane
via a coated pit

Coated vesicle

Fusion of a coated vesicle
with a cisterna

Production of a synaptic
vesicle filled with acetylcholine

Synaptic vesicle fusion
and release of acetylcholine

Binding of acetylcholine
to receptors

Axon terminal

Postsynaptic membrane

Figure 7.6. Depiction of the neurotransmitter release cycle at a typical axon terminal.

ucts to mark the location of the HRP. Then the tissue was killed, and the proteins fixed in place with formaldehyde or other suitable fixative, and the tissue was sectioned and prepared for transmission electron microscopy. The passage of HRP through the axon terminal was traced by varying the time between nerve stimulation and fixation of the tissue.

Heuser and Reese found that when HRP was applied during nerve stimulation, the HRP reaction product first appeared in vesicles that had a distinctive outer coating (Figure 7.6; HRP not shown). After appearing in the coated vesicles, the HRP moved into elongated membrane structures called cisternae, and after that HRP appeared in synaptic vesicles, some of which were positioned for release. Thus, the progress of the HRP product through the system suggested that the HRP had been taken up into coated vesicles of recovered membrane that budded off into the axon terminal, and then fused with the cisternae, and finally new synaptic vesicles constructed of the recovered membrane budded off from the cisternae, filled with neurotransmitter, and moved into position for release. Think of a radial array of conveyor belts where membrane added to the surface of the axon terminal at its center by the fusion of synaptic vesicles flows toward the perimeter of the synapse where it is internalized in coated vesicles. Then the membrane is processed in the cisternae and sent back to the center of the terminal to form new synaptic vesicles and start the cycle all over again.

Not all vesicles in all synapses follow the complete route for recycling. Alternatively, membrane recovered in a coated vesicle may lose its coat and refill with neurotransmitter, becoming a synaptic vesicle without passing through the cisternae. In some synapses the shortcut can be even shorter where the synaptic vesicle fuses with the membrane-releasing neurotransmitter but is not incorporated into the membrane of the terminal. Rather, after releasing its neurotransmitter, the synaptic vesicle backs away and separates from the membrane to refill with neurotransmitter and go on to release again (see Nicholls et al. 2001).

The process of formation of a vesicle from the surface membrane is called endocytosis. Coated pits (Figure 7.6) form as shallow invaginations of membrane into the cytoplasm, which progressively deepen and become spherical as the membrane closes around the opening to the outside and then pinches off within the axon terminal, forming a coated vesicle. The key to vesicle formation lies in the chemical structure of the coat. It takes force to form a coated pit and mold it into a small vesicle from the essentially flat surface of the membrane. The coat protein called clathryn supplies the force. An individual clathryn molecule has three identical arms connected together at the center. The arms are curved, oriented in the same direction, and evenly spaced around the center, creating an overall appearance like a pinwheel. Importantly, the bond angles of the arms cause the tips of the arms to lie in a slightly different plane from the center, resulting in a pinwheel that is slightly dished. Clathryn molecules bind together via their arms, forming a sheet with a pentagonal array. Because the arms of individual clathryn molecules are angled in the same direction and clathryn molecules all bind to each other in the same orientation, the polymer sheet that is formed starts out as a dished surface and as more clathryn molecules are added the sheet begins to look like a geodesic dome and then closes into a little sphere. The clathryn molecules also contain binding sites that bind to specific membrane proteins oriented such that the concave surface of each clathryn molecule faces the surface membrane from the inside. As additional clathryn molecules are added to the perimeter of a growing clathryn sheet, the concave surface of the sheet pulls the membrane into a dish shape, the coated pit, then into a deep spherical invagination still connected to the axon terminal membrane, and then it finally pinches off forming a coated vesicle. In this way the energy of polymerization of clathryn is transferred to the mechanical energy that produces a vesicle from the surface membrane. The polymerization of clathryn is reversible and is regulated by biochemical mechanisms inside the nerve terminal. After a vesicle is formed, the clathryn depolymerizes and the individual clathryn molecules are available to do it all over again.

The choline produced in the synaptic gap by the breakdown of the acetylcholine released from the axon terminals is also recycled. The released acetylcholine is broken down by acetylcholinesterase into choline and acetate (acetic acid). The choline in the synaptic gap is transported back into the axon terminal by a choline transporter protein in the presynaptic membrane. There the choline is used to resynthesize acetylcholine, which is reloaded into synaptic vesicles to be used again in transmission. Recycling of choline and the ongoing synthesis of acetylcholine appear to be necessary to maintain high-frequency synaptic transmission. In a classic study of synaptic transmission in sympathetic ganglia where acetylcholine is also the neurotransmitter, during an hour of high-frequency stimulation of the incoming nerve, the total amount of acetylcholine in the ganglion was released every ten minutes while transmission through the synapses was undiminished (Birks and MacIntosh 1957).[1]

The mechanisms by which vesicles fuse to and are budded off from membranes are fundamental processes not by any means limited to nerve, muscle, and synaptic transmission. The entire cell membrane is produced and maintained by membrane assembled within the cell and delivered in vesicles that fuse to the surface membrane. During embryonic and postnatal development axons grow in length by as much as one millimeter per day with the required membrane lipids, sodium and potassium channels, and most everything else supplied by vesicle fusion.

The complex mechanisms in the axon terminals are geared to release pulses of acetylcholine into the synaptic gap at frequencies of hundreds per second. Now what happens? The function of acetylcholine is to open the acetylcholine receptor channels in the postsynaptic membrane of the muscle fiber. Acetylcholine receptor channels in skeletal muscle are of a type known as nicotinic acetylcholine receptors and are composed of five protein subunits surrounding a channel through the postsynaptic membrane. The channel is conductive to positive ions. The resting acetylcholine receptor channel opening is too small for ions to pass through but opens wider when acetylcholine binds to the end of the acetylcholine receptor channel that is exposed on the outside of the cell. Two identical protein subunits of the channel, called the alpha subunits, are critical in controlling the gate (see Kandel et al. 2013). One acetylcholine molecule binds to a site between each alpha subunit and its neighboring subunit, and an acetylcholine molecule must bind to both sites to open the channel. Therefore, channel opening requires two acetylcholine molecules. The binding of acetylcholine causes a shift in the orientation of the channel proteins, which widens the channel enough for small positive ions, sodium and potassium, to pass through, but not calcium, which is too large. Since the resting membrane is already highly permeable to potassium, the increase in potassium permeability caused by the opening of acetylcholine receptor channels amounts to just a small percentage of the resting potassium permeability. In the case of sodium, the resting permeability is very low, so the opening of acetylcholine receptor channels causes a much larger percentage increase in sodium permeability, resulting in a net inward current carried by sodium. Thus the membrane depolarizes, moving toward the sodium equilibrium potential.

The operation of acetylcholine receptor channels occurs on the molecular scale, so just like the voltage-gated sodium and potassium channels discussed earlier they are subject to the thermal motions of all the molecules surrounding them. The binding of acetylcholine to the receptor channel is reversible and subject to an equilibrium which is determined by the strength of the bond. An acetylcholine molecule stays bound to its receptor until it receives thermal energy from near-simultaneous collisions of surrounding molecules that are strong enough to break it free.

If someone had asked me before 1976 if it would be possible to observe the opening and closing of a single acetylcholine receptor channel on an oscilloscope or computer screen, I would unhesitatingly have said it would never be possible, but that is exactly what Erwin Neher and Bert Sakmann did. As amazing as microelectrode recording is, the patch recording technique they developed staggers the imagination just for coolness (Neher 1991; Sakmann 1991). Instead of a sharp point, a glass patch clamp electrode has a blunt tip with a hole in the middle surrounded by a smooth glass surface. When pressed against a cell membrane, and a small amount of suction applied inside the electrode, the smooth glass surface adheres tightly to the cell membrane, sealing around the perimeter of a 10-μm^2 patch of membrane covering the hole. The patch recording technique has two variations: the electrical activity of the patch membrane can be recorded with the patch in place on the cell; or the electrode can be pulled away from the cell, breaking the patch free around its outer perimeter and leaving it stretched over the tip of the electrode like a drumhead. In the latter configuration, the electrical properties of the patch membrane can be studied in isolation. Voltage-clamp circuitry is used to set the voltage across the membrane patch at a desired level just as in the squid axon experiments of Hodgkin and Huxley, hence the name patch clamp.

Just like Loewi and Millikan before them, when Neher and Sakmann developed the patch recording technique, they knew exactly what they were looking for and found it. Back before there were any measurements made with intracellular electrodes, Bernard Katz and Ricardo Miledi (1972) had reported that putting acetylcholine in the solution bathing an isolated muscle produced small fluctuations in voltage, that is, electrical noise that could be detected with an extracellular electrode. Since the electrical noise was a response to acetylcholine, it seemed likely that it was produced by the opening and closing of acetylcholine receptor channels, and so it seemed to Neher and Sakmann that with about a 100-fold amplification it might be possible to record the electrical impact of the opening and closing of individual acetylcholine receptor channels. The high electrical resistance of glass microelectrodes limits their sensitivity, but the hole in patch electrodes is much larger than the opening in the tip of a glass microelectrode and consequently has much lower resistance that is within the range needed to observe the electrical activity of individual channels. Since the area of the patch is much smaller than the whole cell, Neher and Sakmann predicted that the number of acetylcholine receptors within the patch would be small enough to monitor their individual opening and closing.

The extracellular surface of the cell membrane patch faces the inside of the patch electrode, so Neher and Sakmann added a low concentration of acetylcholine to the conducting solution in the patch electrode to create a constant low level of activation of the acetylcholine receptor channels in the patch. The first membrane used was from frog muscle, of course. Muscle membrane containing acetylcholine receptor channels was accessed in two ways. At first there was no obvious way to patch record from the postsynaptic membrane where the acetylcholine receptor channels are located because the nerve terminal was in the way. However, when muscle is denervated, newly synthesized acetylcholine receptor channels have nowhere special to go so they distribute over the entire muscle surface. Therefore, Neher and Sakmann initially used denervated frog muscle, which they patch recorded from anywhere on the muscle surface. Later they discovered that slightly stronger treatment with the enzyme collagenase, which was normally used to clean the membrane surface for patch recording, would loosen the nerve terminal, which could then be blown away with a tiny jet of water that does not damage the muscle—thus exposing the postsynaptic membrane for patch recording.

Neher and Sakmann immediately found what they were looking for, and electrical recording from cell membranes moved at once down from the nanoscopic scale to the molecular scale. Voltage-clamped membrane patches exposed to acetylcholine in the patch electrode displayed little square pulses of current that could only represent the opening and closing of individual acetylcholine receptor channels (Figure 7.7; note that the high sensitivity required is reflected by the wave pattern in the drawing representing electrical noise). Can you imagine watching the activities of individual molecules on the computer screen? Fantastic! The shape of the current pulse confirmed that the channels are either fully open or fully closed. By varying the ionic composition of the solution in the patch electrode and in the fluid into which the electrode with the

Single acetylcholine receptor channel currents

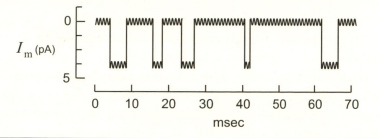

Figure 7.7. Depiction of a patch clamp recording of currents generated by the opening of individual acetylcholine receptor channels. The calibration is picoamps (pA).

excised patch was dipped, they determined that a single open acetylcholine receptor channel at the postsynaptic membrane allowed the passage of potassium and sodium at a rate ranging from 10,000 to 100,000 ions per millisecond.

CHEMICAL TRANSMISSION IS FAST ENOUGH

The process by which neurotransmitters released from the axon terminal cross the synaptic gap is usually described as diffusion. While true, this is somewhat misleading as diffusion connotes a slow process, for example, a droplet of ink spreading out in a glass of water. At long distances, diffusion is extremely slow, but if the distance is short enough diffusion is nearly instantaneous. Membranes are about 8 nm thick.[2] Synaptic vesicles at the vertebrate neuromuscular junction measure about 40 nm across, and the synaptic gap is about 60 nm wide—just 50% greater than the diameter of a synaptic vesicle. Each synaptic vesicle contains about 7,000 molecules of dissolved acetylcholine banging around in constant thermal motion.[3] The arrival of the action potential causes synaptic vesicles to fuse with the membrane of the axon terminal, opening into the synaptic gap. At the initial opening of a vesicle, the straight-line distance across the diameter of the vesicle to the muscle fiber membrane directly across the gap is about 100 nm, but that distance does not persist very long because the vesicle quickly flattens, turning itself inside out, essentially expelling its entire contents into the 60 nm gap. Mathematical modeling indicates that when a vesicle fuses with the axon terminal membrane, the acetylcholine that is released reaches its maximum concentration throughout the gap within 10 microseconds (µsec)—virtually instantaneously!

But the journey of the acetylcholine molecules from the vesicle opening to the acetylcholine receptor channels is perilous because the gap is filled with clusters of acetylcholinesterase molecules tethered to the muscle membrane, like molecular pit bulls on leashes ready to bite them in the acetate, leaving choline, which is unable to bind the receptor. But the good news is that the acetylcholinesterase molecules are outnumbered by about four to one; there are about 10,000 acetylcholine receptor channels per square micrometer of postsynaptic membrane compared to 2,600 per square micrometer of tethered acetylcholinesterase molecules. Also, it takes a tenth of a millisecond (100 msec) for an acetylcholinesterase pit bull to chew up and spit out the remains of an acetylcholine molecule and be ready to do it again, while binding of acetylcholine to the acetylcholine receptor channels is much faster. When a single vesicle fuses and releases its 7,000 molecules of acetylcholine into the synaptic gap, some are immediately grabbed by the pit bulls, but their sacrifice permits the many others to get through and bind to the receptors, and those receptors that have two acetylcholine molecules bound to them open. After a tenth of a millisecond, when the pit bulls are spitting out the remains of their first kills, the vast majority of the surviving acetylcholine molecules are already bound to the acetylcholine receptor channels where the pit bulls can't get them, so the concentration of acetylcholine free in the synaptic gap is low, leaving the pit bulls frustrated and hungry. The bond between acetylcholine and the acetylcholine receptor channels is strong enough so that it takes significant time for the acetylcholine to be released—about 63% of the bound acetylcholine is released each millisecond. By the time the first receptor-bound acetylcholine molecules are released, the pit bulls are ready for them and pick them off as fast as they appear, so each acetylcholine molecule only gets one chance to open an acetylcholine receptor channel and that's it.[4] Thus, the relatively low concentration of acetylcholinesterase allows the full amplitude of the acetylcholine signal to get through while keeping the duration of the signal as short as possible. The result is that transmission frequencies in the hundreds per second are well within the capability of chemical transmission.

The discovery of the mechanism of synaptic transmission provides an excellent example of how a scientist's view of the world affects scientific thinking. Some scientists used to thinking of diffusion as slow had dismissed the chemical hypothesis of synaptic transmission, but at the nanoscopic and molecular scales things happen much faster than is easily imagined. In the end, an assemblage of complex mechanisms were discovered that would have been hard to conceive of in advance, all operating together, to make fast chemical transmission happen.

MECHANISM OF MUSCLE CONTRACTION

The only measure Galvani had to determine if electrical stimulation produced an impulse in the nerve was the resulting muscle contraction, and Alan Hodgkin used muscle contraction centuries later to discover the local circuit at the leading edge of the action potential. No account of animal electricity could be considered complete without a description of the mechanism of muscle contraction.

After the action potential project, Andrew Huxley turned his inventiveness and skillful tinkering to the problem of muscle contraction. As it happened, another British Huxley, Hugh Huxley, originated the modern theory of muscle contraction in the 1950s (Huxley 1953). Andrew made major contributions to supporting this theory based on observations using the interference microscope that he invented. Interference microscopy detects the orderly arrangement of light waves produced when light passes through an orderly arrangement of molecules. Each muscle fiber contains a number of structures called myofibrils that extend the full length of the muscle fiber connecting to the tendons at each end. Andrew Huxley observed patterns of alternating interference bands along a myofibril produced by bundles of short filaments of uniform length made of the polymers of the protein actin alternating with bundles of short filaments of uniform length made of polymers of the protein myosin repeated over and over all along the myofibril. The bundles are arranged end to end along the myofibril with the ends overlapping (see Figure 2.3). Through the work of many scientists, it was shown that muscle contraction is achieved by actin and myosin filaments sliding over one another, increasing the region of overlap and consequently shortening the length of the myofibrils, shortening the muscle, and pulling on the tendons.

The keys to understanding how muscle contraction works are in the structure of actin and myosin. Myosin molecules have a head region connected at an angle by a hinge to a single tail region. The head can swing back and forth. Swinging of the head requires energy and can exert a force in one direction—think of paddling a canoe. The force is applied when the head is pulled back toward the tail. The swinging is regulated, occurring only when the calcium concentration of the intracellular fluid in the muscle fiber is high—another example of movement being regulated by calcium. The tiny forces generated by the swinging of the heads of a great many myosin molecules over extremely short distances are translated into a strong contraction of a muscle by the filament structure of the myofibrils. In a myosin filament the tails lie parallel, forming a rodlike structure, its surface bristling with myosin heads. The myosin filament has mirror-image symmetry with the heads always facing away from the center, so they pull on the filament like in a tug-of-war—think of a myosin filament as two centipedes attached together at their tails. Calcium is the signal that begins the tug-of-war. Actin filaments are also arranged in mirror-image symmetry, but their structure is different. The actin molecules, which have binding sites for the myosin heads, are attached together in a chain, which serves as kind of a railroad track for myosin molecules. Both ends of the myosin filament overlap with several actin filaments. When action potentials in the muscle membrane cause the calcium concentration surrounding the myofibrils to rise, the tug-of-war begins. Each myosin head begins a cycle of swinging forward, binding to the actin filament, pulling backward, releasing from the actin filament, swinging forward, and so forth. The force generated by the myosin heads causes the actin filaments to slide toward the center of the myosin filament, shortening the muscle. Unlike the oars of a scull, the swinging of the myosin heads is not coordinated, so they never all let go at once and the strength of muscle contraction is maintained as long as the calcium concentration is high and the energy supply is not exhausted. This is how all muscles contract. In skeletal muscle and in heart muscle, the alternating actin and myosin filaments are arranged in bundles visible in the light microscope under special circumstances such as interference microscopy used by Andrew Huxley and special staining techniques. It was the banded appearance of skeletal muscle that provided the first clues leading to the sliding filament theory. In smooth muscle, which is distributed throughout the body, such as in the walls of arteries, in the digestive tract, and bladder, the alternation of actin and myosin filaments is present but less obvious.

Unlike the calcium that triggers the release of neurotransmitter, the calcium that triggers contraction of skeletal muscle does not come from the extracellular fluid. Muscle fibers are too big for calcium entering through the surface membrane to reach all of the myofibrils quickly enough and at once to produce a full-strength contraction when demanded by the brain. Instead, calcium ions are contained within a system of internal membranes called the sarcoplasmic reticulum that surrounds the myofibrils along their entire length. In order to trigger the release of calcium, action potentials reach this internal store of calcium by traveling into deep invaginations of the muscle fiber membrane called T-tubules that conduct the action potential to the sarcoplasmic reticulum.

SYNAPSES IN THE BRAIN AND SPINAL CORD

Synaptic transmission is understood in what I will call the "Feynmanian" sense. Physicist Richard Feynman's classic three-volume work *The Feynman Lectures on Physics* (by Feynman, Leighton, and Sands) has been around for half a century and is showing no signs of fading into history. Why? Since the lectures cover the first two years of physics taught at the California Institute of Technology, most of us who revere Feynman lack the math to follow most of what is in them. Their timelessness is not because they teach us physics. In the introductory chapters of the lectures and in Feynman's popular writings, he reveals scientific thinking the way it really works. It is not clear to me why such revelations are so rare, but they are. In one of my favorite Feynman quotations he tells in just a few sentences the nature of scientific understanding: "What do we mean by 'understanding' something? We can imagine that this complicated array of moving things which constitutes 'the world' is something like a great chess game being played by the gods, and we are observers of the game. We do not know what the rules of the game are: all we are allowed to do is to *watch* the playing. Of course, if we watch long enough, we may eventually catch on to a few of the rules. . . . Even if we knew every rule, however, we might not be able to understand why a particular move is made in the game . . . because almost all situations are so enormously complicated that we cannot follow the plays of the game using the rules, much less tell what is going to happen next" (Feynman, Leighton, and Sands 1963, 2.1).

This view of "understanding" is highly relevant to the subject of synaptic transmission. Look just one synapse back from the neuromuscular junction, at the axons in the spinal cord synapsing on the dendrites and cell body of a spinal motor neuron, and the picture becomes enormously complicated because we are entering the central nervous system where the decisions are made. It is across the billions of synapses in the central nervous system where consciousness and everything we think and do are created.

Unlike the muscle fibers, which are its slaves, the spinal motor neuron does not take direct orders from any individual neuron; nor do any of the other neurons in the spinal cord and brain. The dendrites and cell bodies of motor neurons are densely covered with terminals from thousands of presynaptic neurons instead of just one. Unlike the innervation of muscle by the spinal motor neurons, those presynaptic neurons are not just innervating a few hundred postsynaptic neurons. Instead, their axons make thousands of branches and their presynaptic terminals are spread out over thousands of postsynaptic neurons. While action potentials invade all of the axon branches of the presynaptic neurons, the number of synaptic vesicles that release neurotransmitter across each of the synapses is very low, in some cases even a single vesicle per arriving action potential. That single vesicle may contain a very small number of neurotransmitter molecules. For example, in the hippocampus, a major brain region involved in learning and memory, the release of a single vesicle of neurotransmitter may activate only 15–65 receptor channels, which produces a relatively small change in the membrane potential of a postsynaptic neuron and never alone depolarizes it to threshold. So the influence of each neuron in the central nervous system is spread out over thousands of postsynaptic neurons, and it takes the combined action of a great many presynaptic neurons firing at high frequency converging on a single postsynaptic neuron to affect its rate of firing of action potentials. Convergence of many inputs onto a single neuron and divergence of the output of that neuron to a great many postsynaptic neurons are fundamental features of the information-processing system in the brain and spinal cord.

Another fundamental feature of our central information processer is that presynaptic inputs onto a neuron do not all have equal influence. Electrical conduction of a voltage change along a cell membrane that decreases with distance from the source of the current is referred to as decremental conduction. Most neuron-to-neuron synapses are located at a distance from the site of generation of the action potential, so their influence on whether or not the neuron fires an action potential must be transmitted to the firing site by decremental conduction. It is the decremental conduction of a vast array of synaptic inputs onto the neuron that determines whether or not the neuron will fire an action potential.

This is what dendrites are all about. The firing site in most neurons lies at the initial segment of the axon where it connects to the cell body, so an excitatory synapse out on a branch of a dendrite must project its depolarization all the way along the dendrite to the cell body, across the relatively large volume of the cell body, and into the initial segment (Figure 7.8). The voltage-sensitive sodium channels occur at low density in the dendrites and cell body but at much higher density in the membrane of the initial segment, so the decremental depolarization produced by all the active synapses pass through the dendrites and the cell body without opening enough sodium channels to bring their membranes to threshold. Upon reaching the initial segment, the threshold voltage is lower because of the high concentration of sodium channels in the membrane. If enough excitatory synapses fire simultaneously, the resulting depolarization reaches threshold first in the initial segment, which fires the action potential that travels down the axon. Decremental conduction is nondirectional, so the depolarization at the postsynaptic membrane projects in both directions from the synapse along the dendrite and extends into all the branches it encounters, but only the component of the depolarization that reaches the initial segment of the axon influences the likelihood that the neuron will fire an action potential. Synapses that are located on the cell bodies are closer to the initial segment, and thus have a greater influence on the membrane potential at the site of axon generation.

As pointed out in Chapter 6, there are many different neurotransmitters operating in the brain and spinal cord. The major excitatory neurotransmitter is the amino acid glutamate. One of the receptors for glutamate, the AMPA receptor channel, works pretty much like the acetylcholine receptor channel on muscle fibers. Binding of glutamate opens the AMPA receptor channel, which allows the small positive ions sodium and potassium to pass through. The relevant ion is sodium, and the increase in sodium permeability produces an inward current that produces an EPSP.

Inhibitory synapses add an additional complexity to the decision of a neuron to fire or not to fire. As discussed in Chapter 6, gamma-aminobutyric acid (GABA) is the major inhibitory neurotransmitter in the brain. GABA binds to and opens the $GABA_A$ receptor channel, a chloride channel, which produces an inhibitory postsynaptic potential (IPSP). You may recall from Chapter 3 that in squid axon the chloride equilibrium potential at −66 mV is only 4 mV away from the resting potential of −70 mV. While the exact numbers vary between types of neurons, the chloride equilibrium potential is always near the resting potential so an increase in chloride permeability is not going to produce much of an IPSP. In fact, using the squid axon values, we see that the resting potential is 4 mV negative to the chloride equilibrium potential, so an increase in chloride permeability would produce a small inward current resulting in a slightly depolarizing IPSP. Although the slightly depolarizing IPSP would initially move the membrane potential from its resting level toward the threshold for firing an action potential, when EPSPs produced by firing of excitatory synapses cause the membrane to reach a potential that is positive relative to the chloride equilibrium potential, the increased chloride permeability produced by IPSPs will produce an outward chloride current that opposes further movement of membrane potential toward the firing threshold. Thus, an increase in chloride permeability always exerts an inhibitory influence on the firing rate of the neuron. Inhibitory and excitatory synapses on the same neuron fight it out to see who can prevail upon the neuron to fire or not to fire.

Inhibitory synapses located between an excitatory synapse and the initial segment of the axon can short-circuit the current produced at excitatory synapses, resulting in less excitation reaching the initial seg-

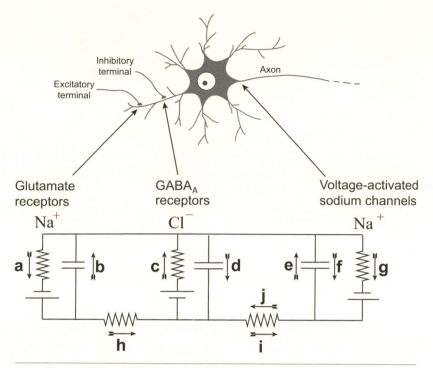

Figure 7.8. Circuit diagram modeling the influence of an excitatory synapse and an inhibitory synapse on the membrane potential at the initial segment of the axon.

ment. To appreciate this, consider the circuit in Figure 7.8. In this example, for simplicity consider the membrane potential to be depolarized to a voltage below threshold, but slightly above the chloride equilibrium potential so that the chloride current (c) is always outward. Opening of the glutamate receptor channels at the excitatory synapse out on the dendrite produces inward sodium current (a) that produces an outward capacitative current (b) that depolarizes the membrane potential at the synapse. Part of the inward current spreads forward through the internal resistance (h, i), producing an outward capacitative current across the membrane at the initial segment (e) that depolarizes the membrane potential toward the threshold. If enough excitatory synapses fire nearly simultaneously, their combined depolarization of the initial segment will cause the membrane potential there to reach threshold, unleashing the inward sodium current (g) that fires the action potential.

Activation of the GABA$_A$ receptor channels at the inhibitory synapse produces outward chloride current (Figure 7.8c), which produces inward capacitative current (d) that has a hyperpolarizing influence on the membrane at the inhibitory synapse. The outward chloride current through the GABA$_A$ receptors also exerts a hyperpolarizing influence on the membrane at the initial segment by drawing internal current away (j), resulting in inward current across the membrane capacitance at the initial segment (f) thereby opposing the firing of an action potential. Additionally, if the glutamate receptor channels and the GABA$_A$ receptor channels are opened by nearly simultaneous firing of the excitatory and inhibitory synapses, some of the depolarizing current extending forward from the glutamate receptor channels (h) will be short-circuited as outward current through the GABA$_A$ receptor channels (c), reducing the forward current (i) that is projected from the excitatory synapse to the initial segment. In this way the excitatory depolarization of the initial segment is reduced by an intervening inhibitory synapse. Thus, in the battle for control of the neuron's decision to fire or not to fire, the location of a synapse is critical. Synapses located in the boonies out on the dendrite tips have far less influence than synapses near or directly on the cell bodies, and the influence of an excitatory synapse can be waylaid by short circuits created by the firing of intervening inhibitory synapses. With thousands of presynaptic neurons synapsing on each postsynaptic neuron the decision to fire or not to fire is complex, indeed.

A motor neuron in the spinal cord is synapsed upon by neurons whose cell bodies are in the brain, spinal cord, and peripheral sensory ganglia. All these inputs bring together the information that will determine if this motor neuron will fire action potentials causing the muscle fibers it innervates to contract. Once the motor neuron fires, the decision has been made, and the muscle has no choice but to follow orders. No single synapse of the thousands converging on the motor neuron can by itself cause the motor neuron to fire an action potential. This is the big difference between synapses in the brain and spinal cord and the neuromuscular junction. As was recognized by Santiago Ramón y Cajal a century ago, the information-processing unit in brain and spinal cord circuitry is the postsynaptic neuron and all the thousands of presynaptic terminals synapsing upon it.

8

USE IT OR LOSE IT

I remember seeing a television program long ago about a woman born with no arms who learned to touch-type with her toes. If you or I tried this it would not work at all. I'll bet you cannot even lift a particular toe on demand independently of the others. Try it—lift just the second toe on your right foot. The circuits in the brain and spinal cord that control your hands are just a lot more extensive and refined than those controlling your toes. Through practice, the woman with no arms developed a similar extensive and refined circuitry in the brain and spinal cord controlling her toes. By practice you might possibly beef up the circuitry needed to develop hands that type at high speed, play the violin or piano, legs and feet that perform modern dance or ballet, and a brain that reads with comprehension much faster than you do now. Of course, talent and body type do have their influence, the younger you start the better, and some goals do require just too much effort to be worth it. But the brain is always capable of change, and a major message of neuroscience is that you have some choice about the direction.

Learning and memory arise from electrically driven changes in the anatomy of the brain. This is perhaps the most important insight gained from neuroscience research during the second half of the twentieth century. The scope of electrically induced anatomical changes extends far beyond what we normally think of as learning. Electrical activity of neurons plays a major role in molding neuronal circuits during the development of the nervous system and during physical activity, even in adults. Much of what you experience and what you do changes the structure of your brain. To see how, we start with the development of the nervous system, where it all begins.

DEVELOPMENT OF THE NERVOUS SYSTEM

A tube that extends the full length of the embryo is the first appearance of the central nervous system during embryological development. A single layer of stem cells lines the inside of the neural tube. Their repeated cell divisions produce the vast majority of the cells of the central nervous system. Some of the daughter cells, as their progeny are called, maintain the stem cell identity and go on to divide and produce more stem cells, while others abandon their stem cell identity and instead go on to differentiate into neurons and a variety of other cell types that are not neurons and are collectively called glial cells or just glia.

The cells destined to become neurons and glia must move from their place of birth in the lining of the neural tube and take up positions at the sites where they will function. Believe it or not, they do this by crawling, using essentially the same crawling mechanism as a primitive single-cell amoeba. Appendages powered by actin and myosin extend forward, attach to the substratum, and pull them along. The mobile cells tend to maintain their positions along the length of the neural tube as they crawl outward, away from its lumen. The first wave forms a second layer of cells just outside the stem cell layer, the next wave crawls through the second layer forming a third, and so forth, building a layered structure that will persist in many regions of the adult nervous system. As the wall of the neural tube thickens, the cells need a little guidance to stay on track as they crawl. To provide this, some of the stem cells differentiate into radial glial cells, which are elongated cells that extend radially from the stem cell layer to the outer surface of the neural tube, like spokes on a wheel, providing the substratum along which the cells migrating outward can crawl. The cell movements quickly get more complicated than this, and at the head end of the neural tube a high rate of cell division and differentiation and larger-scale structural movements ultimately produce a brain.

Some of the neurons in the brain and spinal cord send their axons out into the periphery. Motor neurons in the brain stem and spinal cord send their axons out via cranial and spinal nerves to innervate the skeletal muscles that produce all voluntary movement. Axons that control involuntary responses such as heart rate, blood pressure, and gastrointestinal responses leave the central nervous system to innervate neurons in sympathetic and parasympathetic ganglia, which in turn send their axons to innervate cardiac muscle and smooth muscle. The neural machinery that carries touch and other sensations from the body into the spinal cord and brain develops outside of the neural tube in sensory ganglia. Neurons that develop in the sensory ganglia produce axons that carry sensory information from all over the body into the brain and spinal cord. The neural tube also gives rise to the neural machinery of vision—the retina and optic nerve. The retina is actually a layer of brain cells made sensitive to light and plastered on the back of the eyeball.

Like the initial migration of the cells that become neurons, the connections of axons to the cells that they innervate are established by crawling. A relatively compact cell becomes a neuron by sprouting axons and dendrites. Taking the axon as the example, the growth cone sprouts from the cell and attaches to the substratum. Instead of dragging the cell along as happens during cell migration, the growth cone crawls away from the bulk of the cell, now called the cell body, which contains the nucleus. The growth cone maintains its connection with the cell body via a cell membrane–covered tube that will become the axon. As the growth cone moves along the substratum, the axon elongates behind it creating a resemblance to a spider producing a strand of web. The axon contains microtubules all along its length (see Chapter 6), which function as supply lines that deliver the molecules required to assemble the axon as it grows longer. Axons form branches when growth cones divide in two, each laying down a separate branch as the two growth cones move away from each other.

It is by these basic mechanisms that the structures and connections of the nervous system are produced during development, but sometimes the choreography is a little more complicated. My favorite examples are neurons called granule cells that form part of the major circuits in the cortex of the cerebellum. During part of their development, the migrating cell bodies of the granule cells behave like growth cones, laying down axon behind them as they crawl from an upper layer of the cerebellar cortex to a lower level. There must be other variations on the theme of cell migration and axon formation that are currently unknown. It is a major source of wonder how the movement of cells, their differentiation into many different types of neurons, and their formation of connections with one another can be guided to produce a properly functioning brain, spinal cord, and peripheral nervous system. Several of the principles involved are known, but a precise and verified picture of the major mechanisms that control nervous system development will be out of reach for a very long time.

NERVE GROWTH FACTOR

How the correct neural circuits are formed is the fundamental question of neural development. This question is by no means fully answered, but a part of the answer is that growth cones are guided by chemical signals. A major advance was the discovery of a protein called nerve growth factor (NGF, see Chapter 6).

The early to mid-twentieth century was the heyday of experimental embryology, which basically meant transplant experiments; transplant part of a chick embryo from its normal place to another and see how this affects subsequent development. Chicks were the choice because of the accessibility of their embryos that develop in external eggs. One line of transplant experiments produced results that had tremendous influence on all of neuroscience. Like most experiments, the direction that those results led was somewhat different from the original question. In this case the original question was, why are there more spinal motor neurons in the brachial and lumbar regions of the spinal cord than in other regions? The need for more motor neurons at those levels was clear: the brachial region at shoulder level supplies the innervation to the arms (wings in the case of chicks), and the lumbar region located lower down toward the pelvis supplies the innervation of the legs. In between, the muscles in the trunk, which have a more limited repertory of movements, require fewer motor neurons. The question was, how did it get that way? Which came first, the spinal neurons or the limbs? Or did they develop independently somehow "knowing" they were right for each other?

Rita Levi-Montalcini (1909–2012), who had fled Mussolini's fascist Italy, joined the laboratory of Viktor Hamburger (1900–2001), a famous embryologist at Washington University in St. Louis, and the two began a program of experiments addressing this issue (Levi-Montalcini 1986). They transplanted a developing leg bud from one chick embryo to another, attaching it to one side of the recipient embryo between the normally developing wing and leg. Thus, a region of the developing spinal cord that would normally require a relatively low number of spinal motor neurons was confronted with the additional demands of innervating a developing limb. It turned out that the extra limb somehow caused the region of the spinal cord adjacent to it to become populated with more motor neurons, which then innervated the extra limb much as the normally increased populations of motor neurons in the brachial and lumbar regions of the spinal cord sent out axons and innervated the normal limbs. So it looked like the limb produced from the transplant caused more motor neurons to develop in the adjacent level of the spinal cord. This result implied that the same phenomenon was happening in the case of the normal limbs.

These experiments were performed before there was much knowledge of the chemistry involved in embryological development, and the next question asked by Levi-Montalcini and Hamburger aimed to discover whether transplanting any tissue near the spinal cord would lead to more motor neurons, or if the extra limb exuded something specific to limb tissue that was required to produce the effect. To test this, they transplanted sarcoma tumor cells in the trunk region instead of a leg bud to see if a growing tumor would increase the population of motor neurons just as well. It turned out that the tumor did something similar but even more. So much more in fact that something interesting had to be going on. The dramatic effect of the tumor cells was not so much on the spinal motor neurons, but the peripheral neurons in the sympathetic and sensory ganglia that run along the spinal cord were greatly enlarged adjacent to the tumor, and large bundles of nerve fibers emanating from the ganglia innervated the tumor and also grew all over the place. It looked like the tumor cells were secreting a lot of something that the sympathetic and sensory neurons liked very much.

Now some serious chemistry was needed and so a chemist, Stanley Cohen, joined the group. Chemical analysis requires purification of a substance, and the output of each purification step must be checked for the original activity to determine that the substance is still there. It is time consuming to wait for the chick embryos to show an effect, so an assay more convenient than using whole embryos had to be developed. Methods had been developed to keep chick and mammalian tissues alive, so finally frogs were getting a break! Chunks of dorsal root ganglia dissected out of chick embryos could be kept alive in culture medium, which is essentially the appropriate ionic solution with an array of additives. The addition of sarcoma tumor extract to the mix caused a tremendous growth of axons out from the chunk of dorsal root ganglion onto the surface of the culture dish, providing the assay Cohen and Levi-Montalcini used. At the time, the first step in purifying an unknown biological substance sometimes involved the addition of snake venom containing an enzyme that digested nucleic acids (DNA and RNA). The expectation is either that the activity remains after treatment, showing that the axon growth is not produced by a nucleic acid, or it is greatly diminished or eliminated, suggesting that it is produced by a nucleic acid. Neither occurred. The addition of snake venom to sarcoma extract resulted in a much greater increase in axon growth from dorsal root ganglion pieces than treatment with the sarcoma extract alone. Thou shalt not change more than one variable at a time is the most basic commandment of experimentation. Control groups are important, so Cohen and Levi-Montalcini tested snake venom by itself. Astoundingly, the addition of just snake venom to the dorsal root ganglia culture caused a tremendous increase in the growth of axons from the chunks of dorsal root ganglia. Thus, it seemed that there was also an active ingredient in snake venom with the same kind of effect on nerve growth as the substance produced by the sarcoma tumor. Amazingly, the active ingredient turned out to be exactly the same in both, a protein that was named nerve growth factor, or NGF for short. The name reflects the most obvious effect of NGF in the assay, but it is actually a misnomer because NGF has many other important effects on neurons as well.

Why is NGF in sarcoma tumor cells and snake venom? Tumor cells are aberrant cells that do as they please, so this particular tumor happened to secrete large quantities of NGF, possibly just by chance without any specific benefit to the tumor cells. Snake venom is produced in snake salivary glands. With that clue, Levi-Montalcini and Cohen quickly discovered that the salivary glands of male mice produce large amounts of NGF, so that was the source they used for purification of large quantities of NGF. Male mouse salivary glands were the source of commercial NGF used in research for decades until the molecular revolution made possible the production of recombinant NGF from cultured cells.

The action of NGF upon sympathetic neurons seems optimal for the construction of circuitry. NGF stimulates axon growth by binding to receptors on the growth cones so that the growing ends of axons go where the NGF is (Campenot 1977). Thus, the pattern of release of NGF by cells in the environment of growth cones can determine where the axons grow and form connections. You may be surprised to learn that many cells, neurons included, possess biochemical mechanisms for self-destruction. The mechanism is called apoptosis and involves signaling systems inside the cell that trigger the destruction of the DNA. NGF binding to receptors on growth cones and mature axon endings produces signals that travel along the axon in the retrograde direction to the cell body. The arrival of the NGF signal suppresses the mechanism that produces apoptosis in the cell bodies so the neuron survives. The NGF signal also supports the survival of the axons through which it travels by mechanisms not well understood. Axon branches that lose access to their local supply of NGF degenerate, while other axon branches from the same neuron survive if they have continuous access to NGF (Campenot 1982). These effects of NGF on growth and survival of neurons and their axons fit well into the major patterns of neurodevelopment. The neurons initially establish innervation of more cells than they will innervate in the adult. As development proceeds, some of the neurons die by apoptosis and some of the connections made by axons of the surviving neurons degenerate. Thus, by controlling the initial growth of axons followed by the selective death of some neurons and the selective pruning of axons of the surviving neurons, NGF helps establish the adult pattern of innervation.

NGF does not guide the development of most neurons. NGF was the first discovered of a large group of proteins known as neurotrophic factors that promote the survival of neurons and growth of their axons in a wide variety of neuron types. Many neurons in the brain are responsive to a neurotrophic factor closely related to NGF known as brain-derived neurotrophic factor (BDNF). While much still needs to be discovered about the ways in which NGF and the other neurotrophic factors affect their target neurons and the precise mechanisms of the effects, it is generally accepted that a major function of neurotrophic factors is what has become known as retrograde signaling such as described for NGF above, which functions in the development, maintenance, and remodeling of neuronal circuits. The basic unit of retrograde signaling consists of a neuron, muscle fiber, or other cell type, that releases a neurotrophic factor onto the axon terminals that innervate it or onto the growth cones that travel toward it. The subsequent activation of receptors produce local effects upon the axon terminals or growth cones and retrograde effects that travel back through the axon and into the cell body. A major hypothesis has developed that axons of different neurons compete with one another to innervate the same postsynaptic cells, and in some cases the winners of the competition are the axons that are successful in competing for access to a limited supply of a neurotrophic factor. It is certainly not that simple, but competition of axon terminals of different neurons for postsynaptic partners seems to be a ubiquitous occurrence.

This brings us back to the original question posed by Levi-Montalcini and Hamburger: Why are there more motor neurons in spinal cord regions that innervate limbs? The answer is that the developing spinal cord starts out with a uniformly high density of developing neurons along its entire length—enough to supply a limb anywhere one might develop. The presence of a developing limb is required to support all those neurons, and in the nonlimb areas the excess neurons are eliminated by apoptosis. It is as if the innervating neurons compete with each other for a limited supply of muscle fibers to innervate. Competition between axons from different neurons for innervation territory is a major determinant of circuitry in the brain. Animal electricity enters the picture because which axons "win" in the competition to form and maintain circuitry seems in many cases to be determined by the relative electrical activity of the competing neurons.

ELECTRICAL ACTIVITY AND THE DEVELOPMENT
OF THE VISUAL SYSTEM

With 11.5 billion neurons in the human cerebral cortex (Roth and Dicke 2005),[1] what could possibly be learned by monitoring their activity one at a time? As it turned out, a great deal. As with any area of inquiry, the breakthroughs first occur where the observations are most easily made. Of course *easy* is a relative term, and any studies of the brain are, right from the outset, not the easiest to accomplish. It became very clear soon after the development of the electroencephalogram that the EEG was of limited value in studies of brain function. The next step was recording from neurons in the brain one at a time. Glass intracellular microelectrodes are not well suited for recording from the brain as their tips can break when advancing through tissue. They are best at recording from cells that can be individually seen under the microscope. In order to record the occurrence of action potentials in neurons within the brain, extracellular electrodes made of metal are the way to go.

Recording from single neurons in the brain was made possible by using electrodes made by electrolytically sharpening tungsten wires. Tungsten electrodes were developed by David Hubel while at the Walter Reed Army Institute of Research (Hubel 1957). He would go on to Harvard Medical School, where he would collaborate with Torsten Wiesel on studies of the visual system, for which both would share a Nobel Prize in 1981 (see below). The end of a tungsten wire, far stiffer than steel wire, can be sharpened to a strong, fine tip by passing an electric current through the wire while it is immersed in a potassium nitrate solution. The tungsten preferentially erodes from the sides of the immersed end, while the end itself is protected by the release of gas bubbles that produces a meniscus where they depart from the tip. With a little finesse in varying the length of the end of the wire that is immersed, it is possible to produce a gentle slope to a fine tip a fraction of a micrometer in diameter. The final step is insulating the tungsten electrode up to the tip with a coating of lacquer. A tungsten electrode can penetrate the surface of the brain without breaking. When it enters the cortex and gets close to the cell body of a neuron, the electrode detects the voltage arising from extracellular currents produced when the neuron fires an action potential. Sometimes such an electrode can simultaneously detect action potentials from two to several neurons, the action potentials of each identifiable by their different amplitudes, the larger amplitudes recorded from the closer neurons. A tungsten electrode can be used for an hour or longer, recording from different neurons as it is advanced through the cell layers of the cortex.

Mother Nature has to be credited for some cooperation with neuroscientists by placing much of the important business of the brain in the cortex, readily accessible to the tungsten electrodes. Although a tungsten electrode advancing into the brain kills the neurons directly in its path, destroying some circuitry, the loss of a relatively small number of neurons has no effect on the functions of the cortex. Compare this with what may be the most inaccessible part of the nervous system—the peripheral neurons that control peristalsis in the digestive system. The neurons whose firing produces the peristalsis that automatically moves food through your small and large intestines are hidden away between two layers of muscle in the intestinal wall. Even worse, a functioning intestine is a moving target. In comparison, the brain mostly sits still with all those fascinating neurons motionless near the surface just begging to be studied! Where to begin? David Hubel and Torsten Wiesel made the right choice, which was to study vision. Vision is complex but much easier to get a grip on than, for example, learning and memory. The basic question was: How is the picture of the world that we see through our eyes encoded in the brain? (Hubel 1981; Wiesel 1981).

First, a philosophical point is in order. It is unlikely that any description of the processes of vision in the brain will tell you how you consciously experience vision. The ancients believed, and more moderns than you might think still believe, that the real you is an entity, the soul, sitting inside you that is separate from your body. Thus, consciousness becomes a property of that entity, not the body. Therefore, the eyes and brain are just the machinery that delivers vision to the entity that actually does the seeing. That, of course, is not the scientific view, which depends upon observable entities and processes. In the scientific view you are your skin and everything inside (except, of course, for the multitude of bacteria and other inhabitants). How the processes inside your skin create consciousness is unknown. Perhaps we will find out if and when a computer is developed that is far more complex than present computers and unexpectedly begins to display evidence of consciousness. Then we could begin to figure out the cause—maybe there are consciousness circuits that will ultimately be identified. Until then (and maybe forever) I don't think we have any way of recognizing the processes underlying consciousness. Of course, thinking too deeply philosophically can lead into an intellectual black hole. Descartes thought consciousness was an exclusively human attribute—tell that one to your dog! Anyway, be aware that the scientific analysis of vision will tell a lot about how the brain works but not much about consciousness of vision.

The retina is actually an extension of the brain that buds out into the eye during embryological development. The cornea and lens of the eye direct light onto 100 million photoreceptor cells, the rods and cones. Much processing occurs in circuitry within the retina before their output is turned into a pattern of signals that is sent to the brain proper along the axons of about one million retinal ganglion cells. The 100-fold compression immediately tells you that the 100-megapixel picture captured by the sensor array in the eye is not delivered to the brain. What is delivered turns out to be very different indeed.

If I can be indulged one more philosophical digression, I would say that this should be no surprise. The retina is really a part of the brain, and my experience, and I assume yours, is that "I" am looking out of my body through my eyes. It didn't have to be that way. If the machinery that is actually experiencing vision, be it the soul or otherwise, was essentially watching a closed-circuit television screen inside the brain with the eyes functioning as externally mounted cameras, then the placement of the cameras would not necessarily be an integral part of the experience. But the eyes are part of the experience; we are looking at the world out of our eyes. This makes some sense to me, since the retina is not just a sensor but part of the system that processes visual information.

There is a bit of anatomy you need to know to understand the experiments that follow. The optic nerves from both eyes come together to form an X-like structure on the midline of the brain, the optic chiasm, which makes it look like the optic nerves cross over one another and go to opposite sides of the brain, but that is not what is happening. Instead, within the optic chiasm there is a major reorganization of which axons go where, so it is convenient to speak of optic tracts leaving the optic chiasm to distinguish them from the optic nerves that enter the optic chiasm. Within the optic chiasm the axons from each eye in the optic nerves divide into two groups: axons from the left half of the retina and axons from the right half of the retina. The axons from the left half of the retina of both eyes join together to form the optic tract that connects to the left side of the brain, and likewise the axons from the right half of the retina of both eyes join together to form the optic tract that connects to the right side of the brain. There is just one more twist that we must put up with. When an image passes through the cornea and lens of the eye it is reversed left to right (and flipped upside down). This means that the left side of the retina sees the right side of the world and vice versa. What the eye sees is called the visual field, so the end result is that the left side of the visual field is projected to the right side of the brain and vice versa. To summarize: both eyes connect to both sides of the brain and everything you see to the left of a vertical line bisecting your vision is projected onto the right side of your brain, and everything you see to the right of a vertical line bisecting your vision is projected onto the left side of your brain.

After they enter the brain, the ganglion cell axons carrying visual information terminate within deep structures known as the lateral geniculate bodies, one on each side of the brain. Then the lateral geniculate axons project to and terminate on neurons in the primary visual cortex that is located on the back of each side of the brain. Thus, the information from the left visual field goes to the right primary visual cortex and the information from the right visual field goes to the left primary visual cortex. The cortical neurons in the primary visual cortices send their axons to a wide variety of places in the brain including other areas of the cortex where visual information is further processed and integrated with other sensory inputs.

Hubel and Wiesel set up their experimental animals looking at a screen upon which visual stimuli were projected. They used mostly cats and some monkeys, and fortunately anesthesia was available that produced unconsciousness but did not affect processing of visual information through to the level of the primary visual cortex. They discovered several features in common in the responses of the photoreceptor cells, the retinal ganglion cells, the lateral geniculate neurons, and the cortical neurons. All of these cells are arranged in their respective structures preserving the spatial relationships in the visual field. The visual field is mapped out on the photoreceptor cells in the retina, of course, because they are arranged like the pixels in a camera sensor. Thus, the activity of the photoreceptor cells represents a picture of the visual field. The ganglion cells are spread out over the retina and receive input from nearby photoreceptor cells, so the map of the visual field, although processed, is preserved in the activity of the ganglion cells; that is, ganglion cells that are adjacent to one another in the retina receive inputs from rods and cones in adjacent regions of the visual field. One feature of the processing that is evident at the level of the retinal ganglion cells is that the size of the center of the visual field is enlarged at the expense of the periphery. This makes sense since visual acuity is much greater in the center. Likewise, the visual field is also mapped out on the lateral geniculate body and the primary visual cortex; adjacent cells within these structures also "see" adjacent regions of the visual field. Thus, the picture of the visual field is projected all along the path from retina to cortex with additional levels of processing added at each stage.

Neurons all along the pathway to the cortex, starting with the ganglion cells in the retina, function as contrast detectors. When the researchers recorded the rate of ambient firing of action potentials in a particular retinal ganglion cell or lateral geniculate neuron and moved a small spot of projected bright light around randomly on the screen that the anesthetized experimental animal was "looking at," a firing "hotspot" could be identified in the visual field where the light caused a distinct increase in the firing rate of the neuron. The size of the spot of light could then be adjusted to maximize the firing rate. Small-scale movements of the spot of light revealed a doughnut-shaped region around the firing hotspot, called a "surround," where the rate of ambient firing of the neuron is instead reduced. This pattern of firing is referred to as "on-center, off-surround" because light in the center of the visual field of the neuron increases its rate of firing, while light in the immediately surrounding region reduces the rate of firing. An on-center, off-surround organization of the receptive field of a neuron indicates that the photoreceptors in the hotspot in the center activate the neuron via excitatory synapses, while the photoreceptors in the surrounding region deactivate the neuron via inhibitory synapses. A spot of light large enough to illuminate both the center and the surround results in excitation and inhibition that are about equal to each other and cancel out so that the ambient firing rate of the neuron displays little or no change when the receptive field is exposed to uniform illumination.

The spot of light used in an experiment is nothing like the normal visual input to the eye. Now imagine something a little more normal such as a straight boundary of bright illumination moving over the on-center, off-surround receptive field of a neuron. The illumination first impinges on part of the inhibitory surround causing the firing rate of the neuron to slow. Next the light begins to illuminate the excitatory center, and at some point all of the center and only part of the surround is illuminated, so the firing rate of the neuron is increased. Finally as the illumination covers the remaining segment of the inhibitory surround, both the center and surround are completely illuminated, and the firing rate of the neuron returns to about the same level as it displayed before the boundary of bright light started to pass over its receptive field. Thus, the center-surround organizations of the retina and lateral geniculate function as edge detectors. When not illuminated or completely illuminated, the ambient firing rate of the neuron is about the same, but the neuron changes its firing rate when a change in illumination transects the receptive field. There are also off-center, on-surround ganglion cells and lateral geniculate neurons that behave just the opposite—reducing their firing rate in response to a light-dark boundary.

When Hubel and Wiesel advanced their tungsten electrodes through the layers of the cortex and recorded the action potentials produced in the cortical neurons by the visual stimuli, they found cells that increased their firing rate in response to elongated bars of light of a particular orientation in the visual field. Inhibitory regions flanked the bar-shaped hotspots. These are called simple cells, and other neurons called complex cells were found that responded to more complicated shapes and movement of light across the receptive field. Thus, the transition from the lateral geniculate to the cortex involves a tremendous increase of complexity of processing of visual information supported by a tremendous increase in the complexity of the neuronal circuitry.

This description just hints at the complexity that underlies the processes of vision.

It seems amazing that such a complex system can be produced during embryonic and postnatal development. To begin to understand how, David Hubel, Torsten Wiesel, and their colleague Simon LeVay focused on a simple distinction—how it is decided which cells in the cortex receive input from the left eye and which receive input from the right eye (Hubel, Wiesel, and LeVay 1976). Part of the motivation for these experiments was a direct medical connection. Cataracts in children produce a visual defect that can be permanent, persisting after the cataracts are removed. Since the visual system is wired at birth, this suggests that some of the connections present at birth are degraded if vision is compromised in early childhood, which implies that the electrical activity of the neurons must be maintained for their connections to survive—use it or lose it.

Both the lateral geniculate and the primary visual cortex have six layers of neurons. In the lateral geniculate, each layer of neurons is innervated by ganglion cell axons from the left eye or the right eye, but not both. The segregation of inputs from the two eyes is maintained in layer IV of the primary visual cortex (cortical layers are customarily labeled with Roman numerals), but the circuits from both eyes intermingle in the cortical layers above and below layer IV. When a tungsten electrode is inserted perpendicular to the surface of the primary visual cortex, and a region of the visual field is identified where a light stimulus causes a change in the firing rate, neurons whose firing rate is altered by light in both the left and right eyes are detected in all the layers except layer IV where the cells typically respond to stimuli presented to either the left eye or the right eye, but not both. If the cells in layer IV are responsive to, let's say, only the left eye, then withdrawing the electrode and reinserting it about a half millimeter away from the original site sometimes results in the detection of cells in layer IV that are responsive to light stimulation of only the right eye; an alternation between left and right eye innervation in layer IV occurs on about a half-millimeter scale.

Hubel, Wiesel, and LeVay asked a simple question: What happens to the inputs from the lateral geniculate in layer IV if one of the eyes of an animal is deprived of patterned vision during postnatal development? Deprivation of patterned vision (but not of low-level stimulation by ambient light) was accomplished by suturing closed one eye in the experimental animal shortly after birth. After a period of visual deprivation of one eye it became a lot harder for the neuroscientists to find cells in layer IV that responded to visual stimulation of the deprived eye. Instead, the neurons that had previously been innervated by the deprived eye responded to visual stimulation of the nondeprived eye, which seemed to have taken over the innervation.

So far so good, but it would be very nice be able to grasp what the pattern of innervation actually looks like, and a major technical advance made it possible to look, all at once, at the innervation from both eyes in layer IV over the entire visual cortex. Cells synthesize proteins from amino acids, and a standard technique for labeling proteins during their synthesis is to provide cells with radioactive amino acids, which results in the synthesis of radioactive proteins. A radioactive amino acid commonly used is proline labeled with tritium, which is an isotope of hydrogen: ^{3}[H]-proline (pronounced "H-three proline"). The nucleus of tritium contains two neutrons in addition to the proton that forms the nucleus of an ordinary hydrogen atom. Tritium undergoes radioactive decay with a half-life of about 12.32 years by releasing a low energy beta particle. The decay can be detected by exposure of slices of the cortex to photographic film. When ^{3}[H]-proline is injected into the eye, some of it is taken up by the ganglion cells in the retina and used to synthesize protein. Some of the labeled protein is then transported along the axons of the retinal ganglion cells through the optic nerve and optic tract and into the axon endings that innervate the neurons in the lateral geniculate body. This is not surprising since the vast majority of the proteins in the axon are synthesized in the cell body and transported throughout the axon by the microtubule-based axonal transport system (see Chapter 6). What is surprising is that some labeled proteins are released from the axon terminals of the ganglion cells in the lateral geniculate body and are taken up by the geniculate neurons upon which the axon terminals have synapsed. Then the labeled proteins that have been taken up are transported along the axons of the geniculate neurons into their terminals in layer IV of the primary visual cortex where the labeled geniculate axons make synapses. It turns out that enough labeled protein reaches layer IV to be detected in a slice of preserved cortex exposed against a sheet of photographic film. This sequence of events is amazing. Hardly anyone would have predicted that a labeled protein would cross a synapse like this, but it does. Since the inputs from both eyes remain separate in the lateral geniculate and synapse on separate lateral geniculate neurons, labeled proteins from the left eye will label the circuit from left eye all the way to its destination in the primary visual cortex, and labeled proteins from the right eye will label the circuit from right eye all the way to its destination in the primary visual cortex.

The pictures of labeling in the primary visual cortex after injection of ^{3}[H]-proline into one eye were drop-dead beautiful by any standard (Figure 8.1). Regardless of which eye was injected, the cortex of normal animals displayed zebra stripes with bright labeled regions indicating innervation from the injected eye interspersed between dark unlabeled regions innervated by the uninjected eye, each about a half millimeter wide (Figure 8.1, top).

When one eye was previously deprived of patterned vision during the early postnatal period, and the nondeprived eye was injected with ^{3}[H]-proline, the labeled stripes indicating innervation by the nondeprived eye widened (Figure 8.1, middle) into the regions that would have been innervated by the other eye had it not been deprived. When the deprived eye was injected with ^{3}[H]-proline, the labeled stripes narrowed (8.1, bottom), indicating that the innervation from the deprived eye had withdrawn from the regions now occupied by innervation from the nondeprived eye. (Note that the large apparent holes in the cortex, one at the top of 8.1 and two at the bottom of 8.1, occur because the cortex is not flat, and the slices made through the cortex missed layer IV in the regions of the apparent holes.)

When both eyes were deprived of patterned vision one might have expected that stripes representing inputs from both eyes would become narrow or disappear completely, but that did not occur. Although the pattern of innervation was not completely normal, the stripes from both eyes were each maintained at about a half millimeter in width. Thus, it appeared that when one eye was visually deprived in the early postnatal period, the axon terminals in the cortex activated by the nondeprived eye expanded, displacing the terminals from the inactive circuit connected to the deprived eye. This experiment is a classic example of what has become an established principle of neuroscience—that electrical

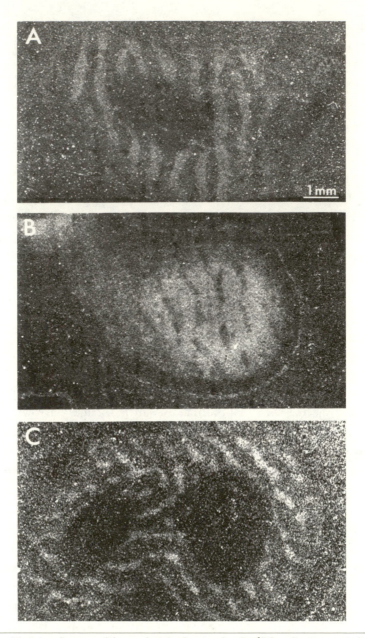

Figure 8.1. Autoradiograms of the monkey visual cortex showing ^{3}[H]-proline transported after injection into one eye: (top) visual cortex of a control animal; (middle) an animal in which the noninjected eye had been previously deprived of patterned vision; and (bottom) an animal in which the injected eye had been previously deprived of patterned vision. (D. H. Hubel, T. N. Wiesel, and S. LeVay, "Functional architecture of area 17 in normal and monocularly deprived macaque monkeys"; reproduced with permission from Cold Spring Harbor Laboratory Press: *Cold Spring Harbor Symposia on Quantitative Biology* 40:581–89, copyright © 1976 by Cold Spring Harbor Laboratory Press.)

activity of neurons exerts a strong influence on the connectivity of the nervous system, with synapses from active neurons displacing synapses from inactive neurons.

"SLICING" INTO THE BRAINS OF LIVE HUMANS

Animal experiments tell us a great deal, but it is always important, circumstances permitting, to make observations in humans. The development of functional magnetic resonance imaging (functional MRI, or fMRI) has revolutionized neuroscience by allowing detection and localization of neural activity inside the brains of humans without the insertion of electrodes, the use of radioactivity, or any other invasive procedure. Functional MRI is one of several techniques that produces images that look like slices through the brain in which the internal structures can be visualized. The very strong magnetic fields of the fMRI machine cause the magnetic fields of the hydrogen nuclei in water molecules to align, which produces a detectable signal.

The strength of the alignment depends upon other molecules in the solution. Functional MRI cannot directly detect active neurons, but it can detect blood that contains oxygenated hemoglobin in the capillaries that supply oxygen throughout the brain. Neurons are major consumers of oxygen, mostly to maintain the electrical potential across their membranes. When neurons become active, the entry of sodium and exit of potassium create a huge demand for energy to run the sodium-potassium exchange pumps to maintain the internal sodium and potassium concentrations at constant levels. During high levels of activity, the capillaries in the vicinity of the active neurons dilate, increasing the supply of oxygenated blood and giving away their positions in the brain to the fMRI machine. The magnetic fields generated by the fMRI machine intersect as they pass through the brain in such a way that the returning signals from oxygenated hemoglobin are limited to a plane transecting the brain, which can be reconstructed by the computer to look like a slice cut through the brain, with the regions of neuronal activity appearing in color against a gray-scale background.

One disadvantage of using increased blood flow as an indirect measure of neuronal activity is that this response requires sustained activity of the neurons to develop; a region of the brain must be active for at least several seconds in order for the activity to be detected. Another disadvantage is that fMRI requires that the subject be inside the enclosed space of an fMRI machine. If you could watch your dog run across your field of vision while your brain activity was monitored by the fMRI machine, the MRI data, once processed and displayed, would not show an image of your dog running across your visual cortex because the fMRI machine's response is not fast enough. But fMRI is great at showing what regions of your brain are involved in more sustained activity, such as a light shining on one part of your retina, with a resolution on the order of a millimeter.

The macula is a pigmented spot visible in the center of the retina that marks the location of the fovea—the part of the retina with the highest density of the photoreceptor cells known as cones. Cones are color sensitive, and the cones in the fovea are responsible for high-resolution color vision at the center of focus of the eye. The fovea detects light from a tiny area, about 2 degrees in the center of the visual field, but the representation of the fovea expands to a larger fraction of the primary visual cortex, so shining a light on the fovea excites activity in the neurons occupying about 20 square centimeters of primary visual cortex called the foveal confluence.

In the disease known as macular degeneration, which destroys central vision, the fovea and a large surrounding area of the retina can degenerate, causing loss of the photoreceptor cells, which can destroy as much as 20 degrees of central vision. When this occurs, the circuitry to the primary visual cortex from the degenerated region of the retina is lost, and shining a light on the fovea and surrounding affected area of the retina produces no response at all in the primary visual cortex detectable by fMRI. People with macular degeneration usually adopt a new point of focus on the retina outside of the damaged area. They no longer look straight at an object they want to see. Instead they look away such that the image of the object is projected on an unaffected region of the retina, called the preferred retinal locus, that remains connected to the brain. Nancy Kanwisher and her colleagues at the Massachusetts Institute of Technology examined the fMRI responses to visual stimuli in two people who had macular degeneration for over 20 years (Baker et al. 2005) and found that light shining on their preferred retinal locus caused the cortex in the foveal confluence region of the visual cortex to become activated in the fMRI image. The only way this could happen is by the formation of new connections that established new circuits between the surviving photoreceptor cells in the preferred retinal locus and the neurons in the foveal confluence region of the visual cortex that had been disconnected from the retina when the photoreceptor receptor cells in the fovea degenerated. The distance required for growing axons to traverse in the cortex to connect to the denervated region is far longer than normal lateral connections made by cortical neurons, and it is not known where in the visual circuits the new connections are made. But clearly, sprouting of axon branches that grew considerable distances had to be involved. It is best to think of the brain as a dynamically changing structure where new connections are constantly being produced and existing connections modified or eliminated as a result of the electrical activity of neurons that arises from experience, movement, and all that goes on in between.

THE "MEANING" OF CORTICAL CIRCUITRY

For a period of a month or two after our dog of 18 years died, when my peripheral vision detected movement I would sometimes actually have the experience of seeing her and turn my head rapidly, half expecting her to come into clear focus on my foveae. Of course she never did, and eventually that experience faded. Although I am a member of the subculture of the scientifically trained, I can still imagine how someone not in this subculture might think he or she was seeing a ghost. Actually, it was hard for me not to indulge that hope since I missed my dog so much, and like most people I would like to think that my loved ones and I might somehow continue beyond this life. But for now I must side with the scientific view that my experience was likely created by my brain alone. I think we now know enough about the brain to imagine how this could work.

Our picture of the world starts out as an image constructed of pixels of light of varying intensities and colors gathered by a sensor array in each eye consisting of 100 million photoreceptor cells distributed over the retina. The image captured by the 100-megapixel sensor changes drastically as it is analyzed, but one feature of the original picture known as retinotopic organization is preserved; the spatial relationships within the picture of the outside world projected onto the retina are preserved within all six layers of neurons in the lateral geniculate body and all six layers of neurons in the primary visual cortex. So when you look at your dog, the picture of your dog projected onto the center of your retinas becomes a dog-shaped array of active neurons in the center of your primary visual cortex. The picture of your dog on the primary visual cortex is not the same as the pixelated picture on the retinal sensor array. The information from the sensor array converges in neuronal circuits onto neurons that are detectors of contrast, or lines of different orientation, and ultimately shapes and movement. Readers familiar with computer graphics software will recognize that pictures can be coded as pixelated images or as what are called "vector" images. Pixelated images are represented by an array of points, 20 million if the image of your dog walking across your visual field projected on 20 million rods and cones on your retina. Every time the dog took a step, each minuscule fraction of a movement changed the intensity and color of the light impinging on many of those photoreceptors, a great many if your dog is a Dalmatian, not quite so many if she is a black Labrador retriever. Thus, it seems a lot to ask of a brain, even yours with a cortex containing 11.5 billion neurons, to keep track of all the changes in the pixels that are your dog when he runs to catch a Frisbee. Instead the brain, analogous to a computer graphics program, changes the visual representations of your dog into something like vector images. In a computer graphics program a straight line can be represented by values that specify its position, length, and orientation, much less information than the individual positions of the thousands of pixels that might make up the line. Curvy lines and enclosed shapes require more complex mathematical representations, but again nowhere near the information required to specify each pixel. The downside of vector images made with computer graphics programs is that they look diagrammatic, for example, the figures I drew for this book. But the upside is that relatively less information is required to change them; for example, moving a line from one position to another or changing the shape of an object, like your dog as he runs across your visual field, is much simpler in a vector image than selecting and changing each pixel. The brain, being the brain with its 11.5 billion–neuron cortex, undoubtedly has the capability of dealing with much more complex vector images than our computers, so the brain-produced vector images need not look diagrammatic, which of course, they do not. This brings us to the "meaning" of the retinotopic organization of the primary visual cortex. Vector images are produced by establishing relationships between pixels that are adjacent to one another, so for the visual system to produce vector images requires neurons to make connections with their neighbors. This happens at all levels in the visual system from the retina through to the primary visual cortex. Similar principles of cortical organization are evident throughout the brain in both sensory and motor systems.

MEMORY CIRCUITS

Now we turn to the age-old question: What is the physical entity that is a memory? Neuroscientists think of a memory as a circuit in the brain that is activated during recall, either conscious recall as in what is called declarative memory of facts or objects, or unconscious recall as in nondeclarative memory, such as of the moves involved in a complex motor skill. The concept of memory as a circuit was challenged several decades ago when some experiments were published that purported to show that a planaria worm could gain the knowledge of a simple learned response by eating a planaria that had previously learned it! The idea emanating from this observation was that memory might be encoded in a chemical structure such as the amino acid sequence of a protein—imagine a protein encoding the appearance of your dog? Ridicule aside, if true there would have to be a vast revision of how we think about brain function. Many researchers got into the act, and the result did not pass the test of reproducibility. Instead, the idea advanced by one of the pioneers of learning theory, Donald Hebb (1949), that memories are stored in networks of neurons has stood the test of time. Thus, a network of neurons fires when you recognize your dog, which did not fire before you adopted her from the shelter. Logically, this must mean that the transmission across some of the synapses in the circuit has been strengthened. Of course, the dog recognition circuits must be extremely complex and variable since you can recognize your dog from many angles and even when the entire dog is not visible, but this is the brain, and with billions of connections each being capable of functioning in multiple circuits, this complexity is well within its capability. It is unlikely that we will ever find the precise circuits that represent your dog in all of her many manifestations to your retinas, but we know that they must be there.

Actually, dogs have played a major role as research subjects in memory experiments. Russian psychologist Ivan Pavlov (1849–1936) first identified a type of associative learning known as classical conditioning, or Pavlovian conditioning, in experiments with dogs. This is the famous experiment in which Pavlov taught dogs to salivate in response to the ring of a bell. This is how it works: When powdered meat is squirted into a dog's mouth, the dog begins to salivate—no surprise. Now if a bell ring precedes the delivery of meat powder, signaling to the dog that it is forthcoming, soon the dog is salivating to just the sound of the bell—no meat required. The meat is the unconditioned stimulus that causes salivation with no training, and the sound of the bell is the conditioned stimulus that acquires the ability to produce salivation by repeatedly preceding the unconditioned stimulus. Classical conditioning requires the memory of a new association; after the training the sound of the bell causes salivation.

The memory circuit that harbors the association between the bell and the salivation response is tucked away somewhere in the complex brain of the dog. Finding a memory circuit where the synapses involved in learning could be identified and studied required a simpler setting. Among the denizens of the Pacific Ocean off the West Coast of the United States is the sea slug *Aplysia californica*. Adults are big slugs, each one a handful. Like the torpedo ray and the squid, *Aplysia* are a gift from the sea. *Aplysia* had attracted the attention of neuroscientists because of the relative simplicity and accessibility of their nervous systems. Brainless, their nervous system consists of only 20,000 neurons with their cell bodies clustered together in ganglia. But *Aplysia* can still learn very simple things, and the goal was to find some simple neuronal circuits that learn and figure out how they do it.

A small minority of the *Aplysia* neurons have large cell bodies, which are identifiable from one specimen to the next. The cell bodies are easily accessible to glass intracellular microelectrodes, allowing synaptic transmission across the synapses onto these neurons to be studied. In the 1970s a cottage industry of *Aplysia* neuroscientists was busy investigating the circuitry of their nervous system. Since then *Aplysia* has become best known for the investigations by Eric Kandel and his colleagues into the mechanisms of learning (see Kandel et al. 2013). They hypothesized that if memory is the strengthening of transmission though neuronal circuits, then it might be possible to find a synapse in *Aplysia* where this happens during learning. With a little bit of luck, learning would not be buried among the majority of the 20,000 neurons that cannot easily be studied. So they looked among the large neurons for synapses where the transmission changed during learning. They found a circuit consisting of just one synapse connecting a stimulus to a response that undergoes classical conditioning.

Aplysia have three body parts involved in the classical conditioning, the tail, the siphon, and the gill. The gill extends out from the body into the water, which is important to protect, as it is used for breathing. If an *Aplysia* is just sitting on the bottom minding its own business, and the tail is given an electric shock, not surprisingly, the gill is quickly withdrawn into the body; tail shock is the unconditioned stimulus and the strong gill withdrawal is the unconditioned response. Touching the siphon also causes the gill to withdraw, but this is a much weaker response. Kandel and his coworkers found that repeated touches of the siphon, each followed by a tail shock, resulted in classical conditioning in which the touch of the siphon became the conditioned stimulus for a strong gill withdrawal, much like the one produced by the tail shock. So in the classical conditioning of *Aplysia,* the strong gill withdrawal response is analogous to salivation in Pavlov's dog, the tail shock is the unconditioned stimulus analogous to the meat powder, and touching the siphon is the conditioned stimulus analogous to the ring of the bell. Importantly, this is not just sensitization of the gill withdrawal produced by repeated electric shocks—the *Aplysia* did not just become jumpy—because the conditioning did not occur if the shock was given before touching the siphon during training. In the "mind" of the *Aplysia,* the touch of the siphon became the signal that a shock was coming. So, fortunately for scientists and dogs, classical conditioning requires nowhere near the brainpower of a dog.

The gill withdrawal response to touching the siphon involves the simplest possible circuitry; sensory neurons innervating the siphon that detect the touch synapse directly on motor neurons that innervate the muscles that cause the gill to withdraw. Therefore, Kandel and his colleagues reasoned that the strengthening of the gill withdrawal

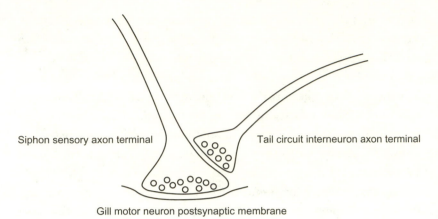

Siphon sensory axon terminal · Tail circuit interneuron axon terminal

Gill motor neuron postsynaptic membrane

Figure 8.2. The arrangement of synapses mediating classical conditioning of the gill withdrawal response in *Aplysia*.

during classical conditioning must involve the strengthening of transmission through the siphon sensory neuron-to-gill motor neuron synapse. For this to work, the tail shock must have a way to strengthen this synapse. The tail sensory neurons, in addition to activating the motor neurons that produce the strong gill withdrawal in response to shock, also activate interneurons that connect by what are called axoaxonal synapses directly onto the axon terminals of the siphon sensory neurons. (Interneurons are neurons that make all of their connections inside the nervous system, that is, they are neither sensory neurons nor motor neurons.) The layout of the two synapses is shown in Figure 8.2. This arrangement suggests that the classical conditioning must be mediated by an increase in the amount of neurotransmitter that is released from the terminals of the siphon sensory axons. Thus there were two questions: How does neurotransmitter release from the tail circuit interneurons produce an increase in neurotransmitter release from the siphon sensory axon terminals, and why does the increase only occur if the siphon sensory axons fire before the interneurons have released neurotransmitter?

Understanding the mechanism takes us into the ridiculously complex world of intracellular signaling. This is the realm of boxes and arrows; this molecule with a complex name (box) affects (arrow) this other molecule with an even more complex name (box), which affects (arrow) . . . You get it. The memorizing-to-understanding ratio in this kind of a treatment is all wrong, so I will skip most of the names. Instead, by way of introduction let me just say that all cells are filled with networks of signaling molecules that pass signals along from one to the next. Some are proteins and some are small molecules. Some signaling inside cells is also mediated by ions. All signaling molecules and ions serve multiple functions, and which functions are activated depends upon their location in the cell and the mechanistic context (that is, what else is happening). The way I see it networks of signaling molecules and internal ions constitute the brain of the cell where the circuits are biochemical rather than electrical, and like the real brain many inputs into the cell connect to outputs by overlapping mechanisms. So I think of the cell as containing a signaling matrix whose complexity is hard to penetrate, but we can learn about how signaling operates during learning in *Aplysia* by considering just six players: the neurotransmitters glutamate and serotonin; calcium; the proteins calmodulin and adenylyl cyclase; and a small internal signaling molecule, cyclic adenosine monophosphate (cyclic AMP).

The neurotransmitter at the siphon sensory neuron–to–gill motor neuron synapse is glutamate, the same neurotransmitter at many of the excitatory synapses in the human brain. Like all neurotransmitters, the trigger for glutamate release at this synapse is the entry of calcium into the axon terminals through voltage-gated calcium channels, which are opened by the depolarization produced by the arriving action potentials. The neurotransmitter at the interneuron-to-siphon sensory axon terminal synapse is serotonin, which is also a neurotransmitter in the brain. The serotonin receptors on the siphon sensory axon terminals are not ion channels. Instead, they directly activate internal signaling molecules.

The memory produced in *Aplysia* during classical conditioning is a long-term increase in the amount of glutamate released by the sensory axon terminals that is produced by the repeated pairing of two convergent signals arriving in the correct order. It works like this. The conditioned stimulus, touching the siphon, activates the siphon sensory neuron-to-gill motor neuron synapse producing a calcium signal in the siphon sensory axon terminals and causing the release of glutamate resulting in a mild gill withdrawal. The calcium also binds to calmodulin, producing calcium/calmodulin that binds to, but does not itself activate the enzyme adenylyl cyclase—a process known as priming. The increased calcium level and the priming of adenylyl cyclase last for only a short time after the siphon is touched. If the conditioned stimulus is followed by a shock to the tail soon enough, serotonin released by the interneurons from the tail circuit binds to receptors on the siphon sensory axon terminal, which triggers mechanisms that activate the primed adenylyl cyclase. Activation of adenylyl cyclase results in the production of cyclic AMP, which activates complex downstream mechanisms that produce a long-term increase the amount of glutamate released by the sensory axon terminals with each action potential. In this way, after many repetitions, touching the siphon produces a strong siphon withdrawal. The order of events is important; in order to strengthen the synapse, the priming of adenylyl cyclase by calcium must precede its activation by serotonin. The reverse order produced by tail shock followed by touching the siphon does not activate adenylyl cyclase. Voilà, in the dim brain of an *Aplysia*, touching the siphon is a signal that the shock is coming.

Get all that? Fortunately, you do not have to memorize it for a test (from me at least!). Think of it this way: A dash of calcium in the active siphon sensory axon terminals, followed by a sprinkling of serotonin on the same terminals resulting from the tail shock that follows, over and over again, increases the release of neurotransmitter by the siphon sensory axon terminals, producing learning in a slug. In this way individual synapses remember that touching the siphon signals that a shock is coming. There are additional learning mechanisms that operate in *Aplysia,* and the molecular mechanisms of learning operating in the brains of humans and animals are more complex and varied. *Aplysia* was a gift of nature, because this animal provided a way to investigate learning in its simplest form within a circuit with only a single synapse. Eric Kandel shared the Nobel Prize in Physiology or Medicine in 2000 for his discoveries about the molecular mechanism of learning.

MEMORIES IN THE BRAIN

In 1966 Terje Lømo, a Ph.D. student working in the laboratory of Per Andersen at the University of Oslo, discovered synapses that remember in the brain (Lømo 2003). The hippocampus, which is buried in the

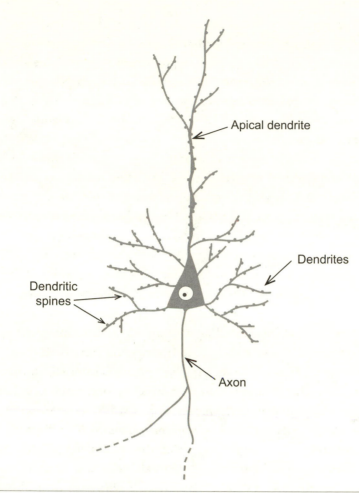

Figure 8.3. A pyramidal cell in the cerebral cortex.

folds of the temporal lobes on each side of the brain, has long been im-
plicated in memory storage because bilateral removal of the hippocampi
severely impairs declarative memory in humans. The principal neurons in the hippocampus and other areas of the cortex have pyramid-
shaped cell bodies which gives them their name—pyramidal cells. A
"typical" pyramidal cell is illustrated in Figure 8.3. There is considerable variation in their structure; readers can find a great many images
of pyramidal cells by searching the Web. Pyramidal cells have one especially long dendrite, called an apical dendrite, which extends upward
from the tip of the pyramid-shaped cell body toward the surface of

the cortex, and many shorter dendrites extending from all over the cell body. The dendrites are studded with small protuberances of their membranes called dendritic spines upon which axon terminals from other neurons form synapses. The receptors for receiving neurotransmitter are localized in the surface membranes of the dendritic spines. A single axon extending from the base of the cell body branches and heads deeper into the brain.

During microelectrode studies of the hippocampus in rabbits, Timothy Bliss and Terje Lømo (1973) found that high frequency stimulation of the synaptic input to neurons in an area of the cortex called the dentate gyrus increased the response of the neurons to transmission across the same synapses for periods of 30 minutes to 10 hours. This phenomenon soon became known as long-term potentiation (LTP), and many researchers joined in, focusing their investigations of LTP on the hippocampus because of its involvement in memory. "Long-term" refers to the duration of LTP relative to synaptic transmission, not relative to the length of memories; LTP is thought to underlie memories that can last minutes to hours but not relatively permanent memories.[2]

There are different kinds of LTP that operate by somewhat different mechanisms. The following discussion, which is based upon LTP in the Schaffer collateral input to the hippocampus, will serve as an example. Schaffer collaterals are axons of presynaptic neurons, each of which branches profusely to form synapses on the apical dendrites of thousands of postsynaptic hippocampal pyramidal cells, and the axons from a great many Schaffer collaterals converge to form synapses on a single pyramidal cell.

Glutamate is the neurotransmitter released at the excitatory synapses involved in LTP. When a presynaptic neuron fires an action potential, the action potential invades all of the branches of its axon, causing glutamate to be released from its axon terminals on all of the pyramidal cells that it innervates. While firing of a single presynaptic neuron releases glutamate on thousands of pyramidal cells, the amount released on each is very small, producing a subthreshold depolarization of the membranes of the postsynaptic neurons. None of the postsynaptic neurons fire an action potential in response to the glutamate released from the synapses of a single presynaptic neuron. It takes the combined effect of glutamate released by a great many presynaptic neurons to depolarize the membrane of the pyramidal cell to threshold, causing it to fire an action potential that travels out its own axon. When the input is insufficient to cause the postsynaptic neuron to fire an action potential, the flow of information stops there. Thus, reaching threshold is the name of the game in information processing in the brain.

The basic neural circuit element that produces associative learning in the hippocampus consists of two excitatory inputs, both converging on a single postsynaptic pyramidal cell. This is different from the *Aplysia* circuit, since there are no inputs on the axon terminals. As in the *Aplysia* circuit, when a weaker input is followed by a stronger input, the weaker input is strengthened. In the *Aplysia* circuit, the two inputs converge on the enzyme adenylyl cyclase, the output of which is cyclic AMP, which triggers mechanisms that strengthen the synapse. In the case of LTP, the two inputs converge upon a special type of glutamate receptor channel, which results in calcium entering the pyramidal cell and triggering the strengthening of the synapse.

There are two different receptor channels for glutamate localized at the excitatory synapses on hippocampal pyramidal cells: AMPA receptors (see Chapter 7) and NMDA receptors. Without going into what these acronyms mean, suffice it to say that these two different glutamate receptors are each named for the different drugs that block the binding of glutamate and prevent activation of the receptor.

AMPA receptors are pretty much standard excitatory neurotransmitter receptors. When glutamate binds to an AMPA receptor, the channel gate opens, allowing sodium and potassium to pass through. The influx of sodium dominates the response since the positive sodium equilibrium potential is far away from the negative resting potential (see Chapter 3). The increased permeability to sodium causes the membrane of the pyramidal cell to depolarize toward the sodium equilibrium potential.

The NMDA receptor channel is the "learning channel." When the NMDA receptor channel opens, calcium flows through in addition to sodium and potassium. Thus, the opening of an NMDA receptor channel allows calcium to enter the pyramidal cell, which triggers the mechanisms that strengthen the synapse. Glutamate alone does not open the NMDA receptor channel because the NMDA receptor channel has two gates: a glutamate-activated gate and a voltage-activated gate, both of which must be opened to allow calcium to pass through. The voltage-activated gate is actually a magnesium ion from the extracellular fluid that binds to a negatively charged site inside the channel acting like a cork that blocks the passage of ions. When the pyramidal cell fires action potentials, the electric field across the membrane produced by the large depolarization repels the positively charged magnesium ion, displacing it from its binding site and ejecting it into the extracellular fluid, opening the gate. Repolarization of the membrane potential after the action potentials have passed allows a magnesium ion to enter and bind inside the channel, plugging it and thus closing the gate.

Suppose there is an array of synapses converging onto a pyramidal cell forming a strong input. When the synapses in the array fire together, enough AMPA receptors open to cause the pyramidal cell membrane to reach threshold and fire an action potential. Suppose there is also an array of synapses on the same pyramidal cell that forms a weak input, and while its firing also may open AMPA receptors, not enough open to depolarize the pyramidal cell membrane to threshold and fire an action potential. Now suppose the firing of the weaker input also opens the glutamate-activated gates of the NMDA receptors at its synapses. This will have no effect on its own if the voltage-activated gates remain shut, but if the stronger input triggers action potentials while the glutamate-activated gates of the NMDA receptors are open, the voltage-activated gates will open, opening the NMDA receptor channels, and calcium will enter the pyramidal cell. The entry of calcium at the sites of the active weaker synapses then triggers mechanisms that strengthen them. Thus, in the case of the synapses on the pyramidal cell, the glutamate signal closely followed by the depolarization-induced calcium signal produces learning.

Earlier in this section I mentioned that synapses on the dendrites of pyramidal cells occur at dendritic spines. These protuberances of the postsynaptic membrane, where the presynaptic axon terminals are located and the glutamate receptors are clustered, create tiny partially enclosed internal spaces where the calcium that enters tends to accumulate. One possible function of the dendritic spines is keeping the calcium concentrated at the synapses through which it entered so that those are the ones that get strengthened.

The calcium entering the pyramidal cell through NMDA receptors strengthens the synapse both by increasing the number of AMPA receptors in the postsynaptic membrane and by increasing the amount of glutamate released by the presynaptic terminals. There is a supply of "extra" AMPA receptors stored in the pyramidal cell within the membranes of small vesicles. Calcium binds to calmodulin and the calcium/calmodulin triggers a chain of events in the pyramidal cell that causes these vesicles to fuse with the postsynaptic membrane, thus adding more AMPA receptors and amplifying the effect of glutamate at the synapse. Silent synapses have been discovered that have only NMDA receptors before learning and only become physiologically active when AMPA receptors have been inserted in their postsynaptic membranes during learning.

Calcium entering the pyramidal cell also activates a mechanism that strengthens the synapse by increasing the amount of neurotransmitter that the presynaptic terminals release onto the pyramidal cell. In order for this to happen, the calcium entering the postsynaptic neuron must have a way of sending a signal back across the synaptic gap to the axon terminals. It is likely that the signal is nitric oxide (see Chapter 6), a small, highly diffusible signaling molecule that is produced when calcium activates the enzyme nitric oxide synthase in the pyramidal cell. The nitric oxide then diffuses back across the synapse where it triggers expansion of existing axon terminals and growth of new ones, resulting in the release of more neurotransmitter onto the postsynaptic neuron when the terminals fire action potentials. The growth of terminals that provide additional release sites is a much longer-term change than the insertion of additional AMPA receptors into the postsynaptic membrane and is believed to mediate longer-lasting memories.

It is a major insight, originally postulated by Santiago Ramón y Cajal and later Donald Hebb, that memories are recorded by the growth of new connections in the brain. It turns out that learning-induced growth can actually be observed in the hippocampus. Research has unequivocally shown that tasks performed by experimental animals that require certain types of memory cause the dendritic spines on hippocampal pyramidal cells to increase in number and size, likely reflecting an increased area of synaptic release sites resulting in increased release of neurotransmitter. Thus, changes in the dendritic spines seem to reflect the participation of the neuron in recording a memory.

Memory is a multifaceted phenomenon, and there is much that remains to be discovered. The idea that memories can be held in the activity of reverberating circuits first proposed by Donald Hebb in 1949 has recently undergone a revival. It appears that high-frequency rhythmic activity called beta and gamma oscillations and low-frequency rhythmic activity called alpha and theta oscillations play roles in maintaining short-term, working memories (Roux and Uhlhaas 2013). While the discussion here has focused on just two examples of memory mechanisms, there are other known mechanisms of learning and undoubtedly more that are yet to be discovered—it's the brain after all, and our ignorance undoubtedly far exceeds our knowledge!

9

BROADCASTING
IN THE VOLUME
CONDUCTOR

Nothing happens in the nervous system or muscle without electric current flowing through cell membranes and spreading into the extracellular space. Unlike wires where current flow is in a linear direction, current spreads throughout volume conductors such as ionic solutions. The extracellular currents produced by electrically active cells form a broadcast into the volume conductor of the body in terrestrial animals like us. Familiar examples are the broadcasts of the heart and brain that reach the surface of the body where they are detected as the electrocardiogram (ECG) and electroencephalogram (EEG). The broadcasting does not stop at the skin in animals that live in water where torpedo rays, electric eels, and electric catfish broadcast potent shocks into the water to stun their prey. In order to understand how such broadcasts are made and their implications and uses we start with the torpedo ray and electric eel, both of which have played major roles in the development of neuroscience.

TORPEDO RAYS AND ELECTRIC EELS

I especially remember two animals from my childhood visits to the Bronx Zoo: the giant panda and the electric eel. I remember the giant panda because of its outlandish appearance and the electric eel because it could light a lineup of 10 or so light bulbs when mildly disturbed by a zookeeper. The panda was a curiosity, but the electric eel was a profound mystery. How could an animal produce electricity? Electric eels don't have batteries and wires in them, so how do they do it? There were no wires connected to the eel so how did the electricity get from the eel to the light bulbs above the tank? In having these thoughts as a child, I was beginning to do what all humans from toddlers to Aristotle and Einstein always do when confronted with a mystery—I tried to relate it to the familiar. I associated electricity with batteries and wires, so where were they? I had no idea that electric fish had been shocking and mystifying humans since the beginning of writing and likely long before that.

The electric eel *(Electrophorus electricus)* lives in the rivers of South America and was unknown to Eurasian civilization before the discovery of the New World. But another dramatically electric fish, the torpedo ray (genus: *Torpedo*) (Figure 9.1), lives in many places throughout the world, including the Mediterranean Sea. Early Europeans had no hope of explaining the shock produced by the torpedo ray since electricity had yet to be discovered and figured out. The explanation had to wait until the second half of the eighteenth century. I am aware of no land animals that produce a strong electric shock. This ability likely evolved in and remained confined to aquatic animals, because they live immersed in a volume conductor. Torpedo rays and electric eels transmit a shock through the water powerful enough to stun their prey. The shocks they produce are the most dramatic examples of electricity projected into the water by aquatic animals. Weaker electrical broadcasting and electroreception are used by a wide variety of aquatic animals for detection of prey and avoidance of collisions with objects they cannot see when swimming through muddy water.

Figure 9.1. Torpedo ray.

For over 1,500 years of recorded history it was anybody's guess what was behind the shock produced by the torpedo ray. Of course, not just anybody's guesses were given credence and written down for future generations. Galen was the first and foremost guesser about medical matters. He classified the ability of the torpedo ray to produce the sensation, now recognized as electric shock, firmly in the category of magic (Finger and Piccolino 2011). Galen described the experience of touching a torpedo ray as a strange feeling of coldness and anesthesia in the hand and arm. This was especially mysterious since the strange feeling could be transmitted from a speared torpedo ray through a trident into the hands and arms of the fisherman. Galen was familiar with poisonous marine animals, but it was hard to see how a poison could act at a distance. Even thinking as long and hard as Galen might have done, without knowledge of electricity he had no hope of understanding how the torpedo shock could be transmitted along the handle of a trident. But then, as now, lack of real understanding did not prevent a phenomenon from being used in medicine. The anesthetic properties of torpedo shock prompted Galen to explore the possibility that it might be used to relieve pain. Galen's investigations led him to conclude that torpedo shock might be useful for treatment of headache, but not hemorrhoids (Copenhaver 1991). One wonders if the patients actually sat on the ray during the testing of the latter.

The shocks of torpedo rays and electric eels were considered such major mysteries that investigations into their nature were carried out concomitantly with the early investigations of the properties of electricity in the eighteenth century. Edward Bancroft, an American scientist working in Guyana, conducted studies of electric eels that he published in 1769. Electric eels are large, as much as 3 feet in length and much more powerful than torpedo rays, packing a painful and dangerous shock, in some cases reaching upward of 600 volts. Bancroft's list of characteristics of the eel shock included transmission through an iron rod and people holding hands, which is entirely consistent with electricity as the cause. Bancroft also included a curious observation, which cannot be right—that the shock can be felt when a hand is held about 5 inches above the water—which is way beyond the capability of the voltages produced by the eel.

Because electric eels are New World creatures they were not readily accessible to the denizens of the Royal Society of London who would go on to prove that the shock is electricity, so most of the work initially focused on torpedo rays, which live in the Mediterranean and in especially great numbers off the Atlantic coast of France. John Walsh, a wealthy British gentleman scientist who also served as a member of Parliament for many years, set out in 1772 under the mentorship of Benjamin Franklin to demonstrate that torpedo rays produce electric shocks. He set up shop in France, first at La Rochelle and then at the Île de Ré. He immediately encountered a fishing boat filled with hundreds of torpedo rays and began work. Unfortunately, torpedo rays, like squid, do not do well in captivity, and few live more than a day after being caught. Most species of torpedo ray produce a relatively modest electrical discharge of about 60 volts, but the shocks produced by the survivors that Walsh studied must have been weaker than torpedo rays can produce when in good health. Walsh tried very hard to demonstrate that torpedo shocks could produce a spark, but a spark was never seen nor heard crossing even a tiny gap between conductors that he created by slicing a piece of tinfoil stuck on top of a layer of sealing wax with a fine knife blade.

Nonetheless, Walsh collected much evidence for the electrical nature of the torpedo shock, which, with the support of Franklin, was well received by the Royal Society. Walsh used a lightly charged Leyden jar to shock fishermen, who reported that it felt just like the shock from a torpedo ray. The shocks from some torpedo rays could only be felt in the fingertip, but torpedo rays that appeared the healthiest could produce stronger shocks felt by eight people holding hands in a chain, and the shocks could be transmitted through a 40-foot wire. It was known at the time that the organs that likely produced the shocks consisted of stacks of cells shaped like hexagonal disks, now called electrocytes. The

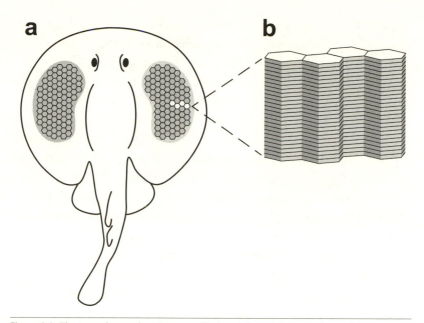

Figure 9.2. The torpedo ray electric organ: (a) dorsal view of the torpedo ray depicting the internal electric organs (electrocytes not to scale); (b) four stacks of electrocytes (just 25 cells are shown in each stack).

electric organs are depicted superimposed on a dorsal view of a torpedo ray in Figure 9.2a where the hexagonal shape of the cell at the top of each stack is visible. Four stacks of cells removed from the organ are depicted enlarged in Figure 9.2b. Each stack contains about 1,000 cells extending from the dorsal to the ventral surfaces of the wings of the torpedo ray, just 25 of which are depicted.

Touching the torpedo ray with both hands on the dorsal or ventral surface did not result in a shock, but a shock was felt if one hand was placed on the dorsal surface and the other on the ventral surface. Two people, one touching the dorsal surface and the other touching the ventral surface, would feel a shock when they touched their free hands together. Walsh found that shocks were not transmitted through glass or sealing wax, which are not conductors of electricity. This last experiment had an interesting wrinkle. Walsh observed that torpedo rays close both eyes just before producing a discharge, so he could tell that the torpedo ray was discharging even though the shock was not transmitted through a nonconductor. Torpedo rays have good reason to close their eyes. The torpedo ray's eyelids likely provide enough electrical insulation so they do not have to see stars or have a brain seizure when they produce a discharge. Certainly nothing known at the time, or discovered since, other than electricity has all the characteristics described by Walsh. He was impressed by the torpedo ray's ability to produce 50 shocks in just 90 seconds, concluding that the electricity is generated internally, and not somehow derived from the surrounding water, prompting him to coin the terms "animal electricity" and "electric organ." But still, there was no spark, and no electrostatic effects of electric charge associated with the torpedo ray were detectable with the instruments available at the time—there was only the sensation produced by the shock to go on. He concluded that the electric charge of the torpedo ray must be too weak. That's the best explanation, but scientists in this kind of position are always holding their breath—even if just a little bit.

If the weakness of the torpedo shock was the problem, the solution was obvious—use the more powerful electric eel to demonstrate the elusive spark. However, electric eels live in the rivers of South America and tended to die during the cruise to England. Nonetheless, before long a few living eels arrived in reasonable shape, and in 1776 John Walsh demonstrated that the discharge of the electric eel generates a spark. After all his efforts, Walsh never published his observation of the spark, although it became widely known that he had finally succeeded in demonstrating a spark produced by an animal.

The electric organs in both torpedo rays and electric eels consist of arrays of disk-shaped electrocytes arranged in stacks (Figure 9.2), which inspired Alessandro Volta to create the first human-made battery using a stack of alternating plates of two different metals. The principle behind the stacking turned out to be the same for the electric organ and the voltaic pile, as his battery was called; that is, adding together many smaller voltages in a series circuit. Henry Cavendish (1731–1810), son of Sir Charles, was a gentleman scientist born into the culture of the Royal Society and had many accomplishments, including being the first to measure the gravitational constant that arose from Newtonian theory. In 1776 Cavendish built an artificial torpedo electric organ out of voltaic piles and showed it could produce a shock in water. The demonstration of the reality of animal electricity in electric fishes raised the possibility that nerve and muscle, once thought to be the conduits of animal spirits, are actually conductors of electricity.

A torpedo ray that has decided to zap a fish for dinner or deter an annoying scuba diver generates a volley of action potentials at frequencies up to 100 per second. The action potentials depart from cell bodies of neurons located in a massive electric lobe on each side of the brain and travel down large cables of axons to the electric organs located in each wing of the ray's body (Figure 9.2). The axon endings synapse on the electrocytes. As in skeletal muscle, the activated nerve endings release acetylcholine that binds to and opens acetylcholine receptor channels on the electrocyte postsynaptic membrane, causing an increase in the conductance to sodium and potassium, in turn causing the membrane potential to depolarize, the inside becoming more positive. Unlike skeletal muscle there are no voltage-gated sodium channels to make action potentials, so the production of the positive-going postsynaptic potential is the only electrical response of the electrocyte to the arrival of action potentials at the axon endings.

Compared to the neuromuscular junction on skeletal muscle, the axon terminals supplying the electrocyte synapse are huge; they cover the entire ventral surfaces of the electrocytes. Thus, the entire ventral membrane of the electrocyte is postsynaptic membrane, crammed with acetylcholine receptor channels. Action potentials arriving at the nerve terminals deliver acetylcholine simultaneously to the entire ventral surfaces of all the electrocytes. Acetylcholine receptor channels are opened and sodium and potassium conductance increases throughout the ventral surfaces of all the electrocytes.

The dorsal surface membrane of the electrocyte has its own specialization. Instead of acetylcholine receptor channels, the dorsal membrane contains a high density of chloride channels. The chloride channels have no gates and so are always open, giving chloride the controlling influence in determining the resting potential, which at −70 millivolts (mV) is very near the chloride equilibrium potential. In effect, the preponderance of chloride channels clamp the dorsal membrane at −70 mV.

In order for an electric organ to produce a high voltage, the electrocytes in a stack must be electrically connected together so that current produced by the membranes of one electrocyte flows through the membranes of its neighbors, like batteries connected in series, rather than short-circuiting between the extracellular spaces separating the electrocytes. Several factors likely contribute to preventing a short circuit. The high density of chloride channels in the dorsal membranes gives them a relatively low resistance, and the high density of acetylcholine receptor channels in the ventral membranes gives them a low resistance when the channels are open, which is when it counts. Thus, the current path between adjacent ventral and dorsal membranes has a low resistance when the electrocytes fire. It is likely that surrounding fat impedes the escape of current from between the electrocytes, giving the current path around the electrocytes an especially high resistance. When a postsynaptic potential is produced simultaneously in all the electrocytes in a stack, each electrocyte behaves like a little battery with the extracellular fluid on the dorsal surface acting as the positive terminal of the battery and the extracellular fluid on the ventral surface acting as the negative terminal of the battery. With each electrocyte connected to its dorsal and ventral neighbors by a low-resistance pathway, the voltages generated by all the electrocytes in a stack add together to produce a strong shock delivered into the surrounding seawater.

The key to understanding how this works lies in understanding how the release of acetylcholine from the axon terminals on the ventral membrane of the electrocyte depolarizes the ventral membrane but not the dorsal membrane, turning a single electrocyte into a battery with the extracellular space on the dorsal surface of the electrocyte acting as the positive pole and the extracellular space on the ventral surface of the electrocyte acting as the negative pole. The extracellular spaces on the different sides of the electrocyte are insulated from one another by fat surrounding the column of electrocytes, which allows them to have different voltages. Voltage is always relative, and the ventral extracellular space in the single electrocytes depicted in Figure 9.3a and b are set at zero. In the column of electrocytes depicted in 9.3c, the ventral extracellular space at the bottom of the column is set to zero. As we proceed from ventral to dorsal, the change in voltage across each membrane that is encountered is indicated at the left.

As we have seen, the cell membrane generates a voltage across itself by virtue of its selective ionic permeability. The selective ionic permeability of the membrane of the electrocyte to potassium and chloride at rest generates a resting potential with the inside of the electrocyte −70 mV negative relative to the outside of the cell (Figure 9.3a). During the postsynaptic potential, the ventral membrane is depolarized to about +10 mV (Bennett, Wurzel, and Grundfest 1961), so in Figure 9.3b

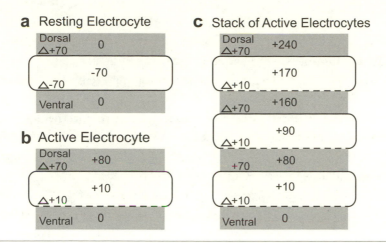

Figure 9.3. Torpedo ray electrocytes showing the voltages that sum to produce an electric shock: (a) electrocyte at rest; (b) active electrocyte during the crest of the postsynaptic potential; (c) a stack of three active electrocytes during the crest of the postsynaptic potential.

the ventral extracellular space is set at 0 mV, and on the other side of the ventral membrane the voltage rises to +10 mV inside the electrocyte. The dorsal membrane of the electrocyte, which is not involved in the synapse, still generates a potential of −70 mV across itself. To achieve this when the voltage inside the electrocyte is +10 mV, the voltage of the dorsal extracellular space must have a value of +80 mV. Thus, when the synapse fires, the electrocyte becomes a battery with the ventral extracellular space functioning as one pole set at 0 mV, and the dorsal extracellular space functioning as the other pole at +80 mV.

You might think it is a bit of trouble for the dorsal membrane to change the voltage of the dorsal extracellular compartment of the electrocyte from 0 mV at rest to +80 mV at the crest of the postsynaptic potential across the ventral membrane. But the dorsal membrane creates a voltage of −70 mV (inside negative) across itself at all times, so riding the crest of the postsynaptic potential is just a natural consequence, instantaneous, and no extra trouble at all. In fact, this is how all batteries behave; if you stack two 1.5-volt AA batteries together in series, 3 volts will instantaneously appear between the terminals at each end. At the crest of the postsynaptic potential the ventral membrane behaves as a 10-mV battery connected in series to the dorsal membrane, which behaves as a 70-mV battery. Putting both together, the electrocyte becomes an 80-mV battery at the crest of the postsynaptic potential (Figure 9.3b). For this to work it is essential for the dorsal and ventral extracellular compartments to be separate; any appreciable extracellular current flowing between them will short-circuit the gain in voltage.

Now as a final step let's connect three electrocytes together in series using the extracellular compartments as terminals (Figure 9.3c). When a postsynaptic potential occurs simultaneously in the ventral membranes of all three electrocytes, each electrocyte adds +80 mV to the amplitude yielding a voltage of +240 mV for the stack at the crest. There can be 1,000 electrocytes in a single stack, each one adding another +80 mV to the total voltage, yielding a total of 80 volts. Since the torpedo ray lives in the ocean, the extracellular spaces at the dorsal surface of the top electrocyte and at the ventral surface of the bottom electrocyte in a stack are connected by seawater. That is where the extracellular current completes the circuit, so the current flows through the column of electrocytes, exits out the dorsal surface of the torpedo ray, flows through the seawater, and then reenters the column of electrocytes through the ventral surface of the torpedo ray.

In this analysis I have assumed maximal values for the resting potential and postsynaptic potential amplitude, no current leakage between electrocytes, and exactly simultaneous firing of all the synapses, so the actual gain in voltage along the stack of electrocytes in a real torpedo ray is likely to be somewhat less. Nonetheless, in the torpedo electric organ some stacks contain more than 1,000 electrocytes connected in series, and each of the two electric organs in a torpedo ray has between 500 and 1,000 stacks. This is consistent with the electrical capabilities of torpedo rays, which produce about 60 volts when the electric organs fire. The current produced by each stack of electrocytes adds together, yielding an impressive output: about 10 amps per electric organ, 20 amps per torpedo ray. Ordinary household circuits put out a maximum of 15 amps at 120 volts, so 20 amps at 60 volts is not too bad for a fish!

Electric rays are all marine animals, but electric eels need to generate higher voltages to overcome the higher electrical resistance of freshwater. Electric eels accomplish this in two ways: The columns of eel electrocytes are horizontal instead of vertical, extending much of the length of the eel's long body and contain many more electrocytes than the columns in the torpedo ray. In addition, the eel uses the action potential mechanism, which generates a higher voltage across the ventral membrane than generated by a synaptic potential. Voltage-gated sodium channels are concentrated on the caudal, that is, tail-facing, surfaces of the eel electrocytes. Opening voltage-gated sodium channels generates a larger change in membrane potential than opening acetylcholine receptor channels. The result is that the head end of the eel can exceed 600 volts positive to the tail end when the electrocytes fire.

Now what goes on in the seawater when the torpedo ray fires its electric organ? Consider the peak voltage of 60 volts. The electric organs span the entire dorsoventral thickness of the ray and are covered on both surfaces by thin skin with little electrical resistance (see Ashcroft 2012). For the analysis, consider the exposed ends of the electric organ as points, like the poles of a battery. It is helpful to speak of two regions of opposite charge between which current flows as a "dipole." The torpedo ray constitutes a dipole, and current flows from the positive pole, known as the "source," through the seawater and to the negative pole, known as the "sink." The current departs the positive pole, spreads out through the seawater, and then converges upon and enters the negative pole. The current takes every available path when flowing through the seawater and is distributed among the available paths in proportion to their relative resistance. A path with twice the resistance of another path will carry half the current. Since resistance increases with the path length, paths that are very long may carry negligible, but never actually zero, current. Thus, theoretically, when a torpedo ray fires off the coast of California, a tiny immeasurable shock should arrive in Japanese waters! But fish and scuba divers more than a few feet away have nothing to worry about, because most of the current density is concentrated very close to the torpedo ray. It is an interesting question why the torpedo ray does not have to worry about zapping itself. The skin of the torpedo ray away from the surface ends of the electric organs, including the eyelids closed over the eyes, apparently provides a high enough electrical resistance to prevent too much current from flowing back into the body of the torpedo ray where it could cause problems for its heart and brain.

The apparent anomaly of the torpedo ray and electric eel puzzled Charles Darwin (1809–1882). "Puzzled" is perhaps not a strong enough word. Darwin knew nothing about the mechanism of generation of electricity by electric fish or about the electrical mechanisms responsible for nerve conduction and muscle contraction; that knowledge would not be available until the next century. But his theory of evolution by natural selection postulated that major evolutionary change occurred by the accumulation of small changes over vast periods of time, each providing a survival advantage, and the ability of a few species of fish to produce strong electric shocks seemed to have arrived out of nowhere. However, the production of electric shocks by torpedo rays and electric eels turned out not to be an anomaly after all. Oceans and rivers are alive with electricity produced by their animal inhabitants and utilized much as the inhabitants of land and air use vision. *Gymnarchus*, the knife fish, so named because of its tapered body fringed by a knife-edge-like ventral fin, belongs to the same family as the electric eel. (Electric eels are not actually related to other eels, but are actually elongated fish.) The knife fish generates current in a sinusoidal wave that projects into the murky water in which it lives. The current returns to stimulate an array of receptors distributed over the fish's head and body. Any objects in the water that are in the path of the current cause a disturbance in the returning current that is detected by the receptors, indicating to the knife fish the location of the object and something about its identity (see Pereira, Aguilera, and Caputi 2012). A nonconducting object that partially blocks the return current may be a rock. Since the knife fish lives in freshwater, and the body fluids of freshwater fish contain high concentrations of ions like the rest of us, a highly conducting object that intensifies the return current may very likely be another fish. This kind of system resembles human-made sonar where an object is detected by the return of a sonic signal sent out through the water. Knife fish and other fish that utilize a similar system can also detect the electric signals sent out by one another, so in their environment of murky water they can move around without hitting objects, find each other, and perform many of the activities such as courtship and mating that other animals depend upon vision to accomplish. These fish produce their electricity by a similar mechanism to those of the torpedo ray and electric eel, but fewer electrocytes are involved, and the voltages produced are much smaller and are only sensed by specialized receptor cells on the fish's head and body. Had this knowledge been available to Charles Darwin it would have pleased him greatly. The evolutionary path is clear: from the basic mechanisms of muscle, to cells that broadcast small electric fields, and finally the addition of broadcasting cells connected in series to ultimately produce the strong electric fields of the torpedo ray and electric eel. Thus the path of incremental survival advantage for each step is clear.

Specialized cells are not necessary for ordinary fish to project electric fields into the surrounding water. Just like you and me, muscle activity in fish generates electric current that flows throughout their bodies, but unlike you and me the fish are surrounded by water, not air, which contains ions, so the current does not stop at the skin. Every time a fish moves a muscle, which is all of the time in the case of the heart and the muscles that pump water across the gills, the electric field and current are projected out into the surrounding water (see Ashcroft 2012; Kalmijn 1982). A flounder or other flatfish may think he is safe lying buried in the sandy ocean bottom, but the ocean floor is being cruised by sharks that have receptors, called ampullae of Lorenzini, spread over their faces that detect the electrical signals projecting into the water above a living, breathing fish. Neither vision nor scent is required; the electrical signal prompts the shark to dive blindly into the sand, open-mouthed, clamp down, and it is game over for the flatfish. Sharks care so little about seeing or smelling the fish that they caused tremendous problems for the telecommunications industry by diving into the bottom and clamping down on undersea communication cables, which project electric fields into the surrounding seawater. Adding extra insulation to cables in areas cruised by feeding sharks solved the problem.

THE ELECTROCARDIOGRAM

Augustus Waller (1856–1922) made the first recordings of electrical signals generated by the heart that reach the surface of the body around 1887. When Waller had his subject immerse each hand into separate pails of saltwater, which were connected to a primitive electrometer, he saw a complex wiggle in the line of the recording that occurred in synchrony with the thump-thump-thump of the subject's heart. The exact shape of the wiggle would turn out to be highly informative about the health of the heart. All cells with electrically excitable membranes generate electrical signals that spread throughout the body, but most of these electrical signals are too weak to be detected without electrical amplification, which was not available until after the advent of the vacuum tube in 1904. Waller was able to detect the electrical broadcast of the heartbeat because of the large signal produced by the synchronized firing of prolonged action potentials in the heart muscle.

The smoked drum recorder that would be used by Alan Hodgkin (see Chapter 4) to record muscle contraction was available to Waller, but movement of the mechanical arm to produce a recording requires muscle—literally. There is no way that an unamplified electrical signal produced by an animal, short of the shock of a torpedo ray or electric eel, could budge a mechanical arm. To record the signals from the heart that reach the subject's hands without electrical amplification, Waller used a device known as a capillary electrometer, invented in 1873 by the French physicist Gabriel Lippmann (M'Kendrick 1883; Zywietz 2002). (Lippmann would go on to win the Nobel Prize in Physics for developing a method for color photography.) The detector in the capillary electrometer consisted of a fine glass tube oriented vertically, with mercury filling the bottom half of the tube and dilute sulfuric acid filling the top half. Hydrogen ions and sulfate ions supplied by the sulfuric acid served to increase the conductivity of the water. Wires were inserted at each end: one in the sulfuric acid solution and the other in the mercury. A small electric current passed through the solutions changed the surface tension of the mercury, causing the height of the meniscus at the mercury-acid interface to change. Because of the very small diameter of the tube, small electric currents produced changes in the height of the mercury column that could be observed with a microscope. Changes in height could be amplified and recorded by projecting a bright light across the interface onto a photographic plate. When electrodes were connected through buckets of saltwater to both hands of a subject, and the plate was moved at constant velocity across the beam to create a time scale (as in rotating a smoked drum), the changes in the height of the mercury column were recorded as changes in the shadow cast by the mercury, producing the first recording of the ECG. Astonishingly, this arrangement produced an ECG trace that, although somewhat distorted, was uncannily similar to the ECG traces eventually produced by modern amplifiers.

Although Waller had provided proof of principle for the usefulness of the ECG, the next step toward its perfection fell to a successor, Willem Einthoven (1860–1927), who developed the next generation electrometer during 1900 to 1903. This was the last advance in the technology of the ECG for about 25 years when vacuum tube amplifiers replaced mechanical amplification. That was enough time for Einthoven to establish himself as the principal developer of the electrocardiogram.

The name of Einthoven's device, string galvanometer, is a bit misleading since the "string" is actually a fine quartz fiber with an electrically conductive coating of silver. It is a fundamental principle of physics that electric current flowing through a wire produces a magnetic field that surrounds the wire. The string of the galvanometer is suspended between two electromagnets oriented one on each side along its length. When current passes through the string, the magnetic field emanating from the string interacts with the magnetic fields produced by the electromagnets, causing the string to deflect into a bow shape with the greatest amplitude at the center. The magnitude of the deflection at the center of the string is proportional to the electric current flowing through it. The shadow of a small section of the center of the string was projected through lenses producing an amplified image on a moving photographic plate. A tiny current of 10^{-11} amps that produced a deflection of the string nearly invisible to the naked eye projected a one-millimeter displacement of its shadow on the photographic plate. Achieving even tiny movements like this required an instrument with massive electromagnets flanking the string. But the result produced a much more accurate ECG record than obtained with the capillary electrometer since the response of the string was faster than the movement of the mercury-water boundary, requiring only about 1 millisecond (msec) to respond to a 2-mV change in voltage (Zywietz 2002).

Although vacuum tubes were invented in 1904, considerable development and refinement over the next three decades was required before electrical amplifiers were developed that transformed electrocardiography into its modern form. The result was that ECG signals could be amplified to enough strength to drive mechanical pen recorders, creating a reasonably faithful record of the amplitude and time course of the ECG. A mechanical pen recorder is a more modern version of the smoked drum, with paper moving past pens mounted on mobile arms supplied with ink fed through small plastic tubes. When an oscilloscope was used as the monitoring device instead of a pen recorder, the accuracy of the ECG was near perfection, as are ECGs recorded with modern computer-based equipment.

The electrical signal broadcast by the heart consists of a complex wave that lasts approximately 0.8 seconds (Figure 9.4). First a low bump appears, lasting about a tenth of a second, called the P wave, that is produced by the depolarization of the membranes of the heart muscle cells in the two atria. After the P wave, the trace returns to baseline for about a tenth of a second after which there is a sharp but shallow dip below baseline called the Q wave. A large spike above baseline called the R wave and then another sharp dip slightly below baseline called the S wave follow immediately. The Q, R, and S waves together compose the "QRS complex," which is generated by the depolarization of the two ventricles. The trace returns to baseline for a little over a tenth of a second, and then a low bump appears that is produced by the repolarization of the ventricles. The entire ECG signal lasts about three-quarters of a second. These are the only large signals broadcast by the changes in membrane potentials during a heartbeat. The signals produced by the repolarization of the atria and the signals produced by the firing of action potentials by pacemaker cells and conducting cells within the heart (see below) are too small to be detected by ECG electrodes at the surface of the body.

Of course, the subjects were no longer connected to the equipment by sticking their hands in pails of saltwater all the while these technological advances in recording were being made. Instead, metal electrodes attached to the body through a jelly made conductive by the addition of ions was developed. The modern ECG is recorded with a number of leads, as the electrodes are called, connected to the arms, legs, and across the chest. The ECG waveform is produced by the voltage difference between various combinations of two leads, and its precise shape depends upon the leads from which the recording is made. The legs, both being connected to the torso below the heart are considered electrically equivalent, and the two arms and one leg together form what is called Einthoven's triangle. Standard ECG limb leads I, II, and III mea-

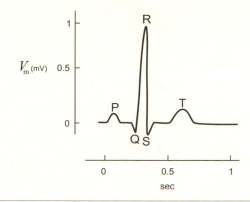

Electrocardiogram

Figure 9.4. Depiction of an ECG record during a single heartbeat.

sure the voltage between the left and right arms, a leg and the right arm, and a leg and the left arm respectively. Since the action potentials in the heart muscle cells pass over the heart in a characteristic pattern with each beat, changes in the pattern of the ECG wave, such as produced by the failure of some heart muscle cells to participate as the result of a heart attack, shows up in the relationship between the ECG recorded at the limb leads and at leads arranged across the chest (Becker 2006).

THE PUMPING HEART

We now believe that the whole point of the heart is, of course, to pump blood; as far as anybody knows the heart doesn't spend any time informing the brain what to do, as was once thought by many, or controlling our emotions, as implied by many modern figures of speech. Although the heart seems relegated to the lowly task of pumping the lowly blood, it accomplishes this task with a certain pizzazz. Unlike mechanical pumps made by humans, which can operate with a few parts that move in relation to one another, such as a piston inside a cylinder, the operating principle of the heart is the generation of squeezing force in muscle surrounding a chamber, thereby reducing its volume and forcing the blood out. The heart has four chambers that operate this way.

The heart is a dual system, with the right side of the heart pumping blood to the lungs where it is oxygenated and then returned to the left side of the heart, which in turn pumps the oxygenated blood throughout the tissues of the body. After unloading its oxygen in the tissues, the blood is returned to the right side of the heart, and the cycle of circulation repeats. The pumping is accomplished on each side of the heart by the atrium that receives blood flowing into the heart, connected directly to the ventricle, which pumps the blood out of the heart. This direction of blood flow results from the placement of one-way valves at the entrance to the atrium and between the atrium and the ventricle, which prevent back flow of blood during contraction. The presence of two chambers on each side of the heart rather than just one provides for efficient pumping because the atria can relax and fill with returning blood while the ventricles are contracting.

When the heart beats, the heart muscle cells in the atria must all contract nearly simultaneously, slightly thereafter the heart muscle cells in the ventricles must all contract nearly simultaneously, and this sequence must be repeated every second or so. The heart must do this all by itself even when there are no instructions or signals of any kind coming from anywhere else in the body. To accomplish this, the heart has a system that automatically generates action potentials that travel throughout the muscle cells in the entire heart to trigger each beat. There are signals sent from the nervous system to the heart via the vagus nerve to slow the heart and via the cardiac accelerator nerve and circulating adrenaline to speed up the heart (see Chapter 6), but they only modulate the rate of the rhythm generated automatically by the heart.

Heart muscle cells are very different from skeletal muscle fibers. Although skeletal muscle fibers are formed by the fusion of many stem cells, each skeletal muscle fiber behaves electrically like a single cell that extends the full length of the muscle. Therefore, the contraction of each muscle fiber independently contributes to the force generated by the muscle. Instead of operating as independent units, the individual heart muscle cells are connected together in electrical and mechanical continuity across the entirety of the atria and ventricles. The connections are distinct structures called "intercalated discs" through which the contractile fibers of actin and myosin of the adjacent heart muscle cells are connected together, end to end, through proteins that extend through the membranes of both adjoining cells. Each heart muscle cell can attach through intercalated discs to three or more neighbors, so the heart muscle cells together form a contractile mesh over the entire surface of each atria and ventricle. When the heart muscle cells of the atria or ventricle contract all at once, the chamber squeezes, pumping the blood.

In addition to linking the contractile fibers of the heart muscle cells together, the intercalated discs link together the electrical activity of the heart muscle cells. The membranes of the two heart muscle cells brought together at an intercalated disc contain ion channel proteins, called connexons (see Chapter 6), that align together to form low-resistance paths for ions that carry electric current across both membranes. Connexons have no gates and so are always open. Since all the heart muscle cells in an atrium or ventricle are connected by intercalated discs, the forward current loops at the leading edge of the action potential pass right through the intercalated discs, so an action potential generated anywhere in the atrium or ventricle will travel over the entire chamber wall.

THE BEAT GOES ON

During the development of the heart, some of the cells become specialized to serve as pacemaker cells or conducting cells. The pacemaker cells generate the automatic rhythm of the heart, and the conducting cells establish highways of especially fast conduction through the heart that cause the muscle fibers in the atria to contract nearly simultaneously followed by the nearly simultaneous contraction of all the muscle fibers in the ventricles.

Action potentials begin in the pacemaker cells in the sinoatrial node (SA node) at the top of the heart. The SA node cells are connected by intercalated discs to conducting cells that in turn make connections that are distributed to heart muscle cells throughout the atria. When the SA node pacemaker cells fire, action potentials rapidly propagate across intercalated discs into the conducting cells and then into the heart muscle cells of the atria, triggering near simultaneous contraction. Meanwhile, conducting cells within the atria also conduct action potentials to a set of cells at the atrioventricular boundary called the atrioventricular node (AV node). Action potentials are then conducted from the AV node into fast conducting cells in the ventricles, called Purkinje fibers, which conduct action potentials to heart muscle cells throughout the ventricles, triggering their near simultaneous contraction to complete the heartbeat. Specialized conducting cells are necessary in the atria and ventricles because the conduction velocity of ordinary heart cells is only about 0.5 m/sec. If action potentials had to travel from the pacemaker directly to the heart cells, the heart cells at the top of the chamber near the pacemaker would contract significantly earlier than the heart cells at the bottom of the chamber. In effect, the conducting cells perform a similar function in the atria and ventricles as the giant axons perform in the squid mantle—producing a simultaneous contraction that squeezes the entire chamber all at once.

The cardiac action potential is somewhat different in pacemaker cells, conducting cells, and heart muscle cells, and several different membrane channels are involved. Fortunately, we do not need to get into the full complexity to understand how the heartbeat is produced. The properties of two types of channels—HCN channels and voltage-activated calcium channels—are key. Together they generate the rhythm of the heart and inject calcium into the heart muscle cells to generate their contraction.

The heartbeat begins with the firing of an action potential in the SA node pacemaker cells. Unlike the sodium-based action potentials in nerve and skeletal muscle, the pacemaker action potential is generated by the opening of voltage-gated calcium channels. The calcium concentration outside the cell is about 2.5 millimolar (mM) and inside the cell is only 0.0001 mM, which yields a Nernst potential (see Chapter 3) of about +134 mV, so the opening of calcium channels produces inward current resulting in a depolarization of the membrane potential just like the opening of sodium channels. Voltage-gated calcium channels have two gates that are analogous to the two gates in voltage-gated sodium channels—a calcium activation gate that opens first in response to depolarization that produces the rising phase of the action potential and a calcium inactivation gate that subsequently closes in response to depolarization producing the falling phase of the action potential back to the resting potential. The calcium action potential in pacemaker cells propagates along the membrane like a sodium action potential, but the kinetics are much slower, with the action potential lasting about a half second instead of 3 msec (Figure 9.5).

Just as in nerve, to fire action potentials the membrane potential of pacemaker cells must be depolarized to the firing threshold. Unlike in

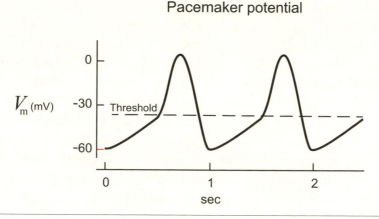

Figure 9.5. Depiction of two action potentials in a cardiac pacemaker cell.

nerve, the heart beats autonomously without external stimulation because the depolarization to firing threshold between heartbeats occurs automatically. The HCN channels produce the automatic depolarization. The opening of HCN channels depolarizes the membrane of pacemaker cells to threshold by allowing sodium and potassium to pass through, similar to the mechanism of depolarization by the opening of acetylcholine receptor channels in muscle (see Chapter 7). However, unlike the acetylcholine receptor channels, the HCN channel gates are voltage activated like the gates in voltage-gated sodium and calcium channels. However, the HCN channel gates are opened by hyperpolarization and closed by depolarization—the reverse of the sodium and calcium channel gates that generate action potentials. The presence of the HCN channels in the membranes of the heart cells makes it impossible for them to maintain a stable resting potential, because repolarization of the membrane after a beat causes the opening of the HCN channels in the pacemaker cells. The result is that at the end of the action potential in heart cells the membrane potential of the pacemaker cells automatically depolarizes to threshold and fires the next action potential (Figure 9.5).

The heartbeat is generated like this. During the action potential, the depolarization closes the HCN channel gates. After the action potential, the membrane repolarizes to about −60 mV, which causes the HCN channels to begin to open. The influx of sodium through the HCN channels causes the membranes of the pacemaker cells to automatically depolarize to threshold, which takes about half a second. Then the pacemaker cells fire action potentials initiating the next heartbeat. During the action potential, the HCN channels close. Once the membrane potential repolarizes after the action potential is complete, the HCN channels begin to reopen, beginning the automatic depolarization to threshold that triggers the next action potential, causing the next heartbeat, and so forth, again and again, your entire life.

The action potential in the heart muscle cells is somewhat different than the action potential in pacemaker cells, because it serves the dual function of propagating the signal to contract over the entire heart and injecting calcium into the heart, which stimulates the contraction of the heart muscle, just as calcium entry into the cytoplasm of skeletal muscle fibers stimulates their contraction (see Chapter 7). This is accomplished by action potentials produced by fast voltage-gated sodium channels and slow L-type voltage-gated calcium channels functioning together. ("L" stands for long-lasting kinetics.) The cardiac action potential rises fast (Figure 9.6) due to the opening of the voltage-gated sodium channels, and the fast rise is followed by a broad depolarized plateau lasting about 300 msec, which is maintained by the slower opening of the calcium channels after the sodium channels have closed. During the prolonged action potential enough calcium is injected into the heart muscle cells to trigger a strong contraction. It is the prolonged action potentials firing nearly simultaneously, first in all the muscle cells in both atria and then in all the muscle cells in both ventricles, that broadcast their currents to the surface of the body where they produce voltage changes recorded as the ECG.

Although automatic, the heartbeat is not beyond the influence of the nervous system, which sends signals for the heartbeat to speed up or slow down by influencing the opening of the HCN channels in pacemaker cells, and thereby determining how fast the automatic depolarization between beats reaches threshold. This is accomplished by the action of neurotransmitters and hormones.

In addition to being activated by voltage, the gate on the HCN channel can be opened by a specific molecule, cyclic AMP, the same small intracellular signaling molecule that is involved in learning (see Chapter 8). Cyclic AMP binds to the channel protein on the inside of the

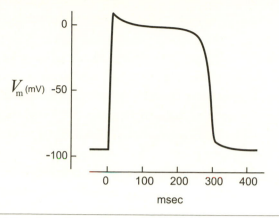

Cardiac muscle action potential

Figure 9.6. Depiction of an action potential in cardiac muscle.

cell. Now we can define the HCN acronym. It stands for <u>h</u>yperpolariza-tion and <u>c</u>yclic <u>n</u>ucleotide-gated. As described in Chapter 3, channel proteins are in constant equilibrium between the open and closed state. Both hyperpolarization of the membrane potential and an increase in the intracellular concentration of cyclic AMP move the equilibrium in the direction favoring the open state of the HCN channel activation gates. Noradrenaline released by the endings of axons of the cardiac ac-celerator nerve onto receptors on the pacemaker cells activate an internal biochemical mechanism that causes cyclic AMP to increase inside the cells. The increased cyclic AMP increases the rate of HCN channels opening between heartbeats. The resulting increase in inward current causes the membrane potential to depolarize to the threshold faster, thus reducing the interval between heartbeats. Adrenaline released into the bloodstream from the adrenal gland also binds to the same receptors as noradrenaline and produces an acceleration of the heartbeat by the same mechanism.

The opposite effect is produced by acetylcholine released by the ter-minals of axons from the vagus nerve. Acetylcholine binds to muscarinic acetylcholine receptors on the pacemaker cells. These are different from nicotinic acetylcholine receptors at the neuromuscular junction and many synapses. Muscarinic acetylcholine receptors are not ion chan-nels. Instead of opening ion channels, binding of acetylcholine activates internal biochemical mechanisms that produce a reduction in cyclic AMP that reduces the number of HCN channels that are opened between heartbeats, reducing the inward current, thus slowing the heart rate.

Since the SA node pacemaker tells the heart when to beat, how can someone survive a malfunction of the SA node? The reason is that HCN channels are also present in the cells in the AV node and in the Purkinje fibers, but at a lower density than in the membranes of the pacemaker cells in the SA node. Under normal conditions, the larger current generated in the membranes of the SA node pacemaker cells between heartbeats causes the membrane to depolarize to threshold first, generating the action potential that propagates everywhere in the heart. When the action potential reaches the AV node cells and Purkinje fibers, which are in the process of producing their own depolarization to threshold, the large depolarization slams their HCN channels shut, so the AV node cells and Purkinje fibers do not get to produce their own action potentials. Once the action potential has passed, the membrane potentials of all the cells in the heart are reset to the resting level, the automatic depolarization begins again in the SA node, the AV node, and the Purkinje fibers, and again the SA node wins the race to threshold, generating the next beat. If the function of the SA node is compromised, it may no longer fire the action potential before the cells of the AV node reach threshold. Because the depolarizing current generated by the HCN channels in the AV node is less, the interval required to reach threshold is longer, and the heart chugs along at a reduced and less coordinated rate. The heart function in this circumstance is reduced, but it is enough to keep a person alive long enough to receive an artificial electronic pacemaker that restores a normal heart rate (see Chapter 10).

BROADCASTS FROM THE BRAIN

The heart is by no means the only broadcaster in the volume conductor of our bodies. All electrically active cells produce broadcasts. Skeletal muscle fibers fire in unison when a muscle contracts, and the broadcast from the muscle action potentials reach the skin surface over the muscle where it can be recorded as the electromyogram (EMG), which has applications in clinical diagnosis and in the control of prosthetic limbs (see Chapter 10). But the most interesting generator of electrical signals to the surface of the body is the brain. When firing in unison in large numbers, neurons in the cortex of the brain produce broadcasts of sufficient strength to pass through the skull (living bone is a conductor) to the surface of the scalp. In 1929 Hans Berger (1873–1941), a member of the faculty of the Friedrich Schiller University at Jena in Germany, was the first to publish an EEG recorded from the surface of the human scalp (Borck 2008). Berger had some far-out ideas. He thought that by measuring blood flow and energy consumption in the brain he could determine a value for the psychic energy that was converted into thought (Millet 2002). But psychic energy (whatever that is) aside, his ideas were somewhat prophetic, since high levels of energy utilization by active neurons resulting in increased blood flow is the basis of modern imaging of brain activity by functional MRI (see Chapter 8).

Observation of the EEG revealed electrical signals reaching the surface of the head with frequencies ranging from 1 to 30 per second at amplitudes of 20–100 mV (see Kandel et al. 2013). The EEG originates mostly from summed synaptic potentials, since nerve action potentials are too brief to transmit well through the skull. The resting activity of the brain forms a wave pattern known as the alpha rhythm, which trucks along at a frequency of about 8–13/sec. Intense mental activity produces faster activity called beta waves that are of lower amplitude and less regular frequency than alpha waves. Alpha waves came as a surprise to Lord Adrian, who thought of the brain as a relay station connecting stimulus to response, which should quiet down when there is no stimulus (Borck 2008).

There was a flurry of interest in utilizing the EEG to reveal how the brain functions, but that goal was never realized. I was in the graduate program in 1968–1971 at the Brain Research Institute at UCLA, where there was great interest in recording from the brain. One of my fellow students commented that discovering how the brain works from electrical recordings from the scalp has about as much chance of success as discovering how a radio, television, or computer works by recording the electrical signals that reach the surfaces of their cabinets. If one ignores the minor detail that not many electrical signals would likely reach the surface of a radio, television, or computer because their cabinets are mostly filled with air, which is not a good conductor, this comment makes a lot of sense. It is now widely accepted in the neuroscience research community that the broadcasts of neurons in the volume conductor of our bodies are not part of the functioning of the brain and the rest of the nervous system. They appear to be epiphenomena; that is, a by-product of the firing of action potentials and synaptic potentials without a function of their own.

But the EEG has been studied for many years and has unquestionable medical applications. Investigations usually begin with the most obvious manifestations, which in the case of the EEG is the electrical storm known as epilepsy. Epileptic seizures begin at some site in the brain and spread to other regions, and if the EEG is being observed with an array of electrodes arranged over the skull, it is sometimes possible to determine where in the brain the seizures begin. In some cases where debilitating seizures cannot be controlled with drugs, the site of origin has been surgically removed with beneficial effect. This approach has been used extensively in temporal lobe epilepsy where recording electrodes are actually inserted through the skull into the temporal lobes of the brain to determine the site of origin of the seizures. Brain tumors are masses of electrically inactive tissue, which can sometimes be localized by recording the EEG, but this use has been supplanted by modern imaging techniques such as magnetic resonance imaging (MRI). The EEG is still used extensively to study the activities of the brain during sleep and to determine brain death in comatose patients.

The EEG never fulfilled the promise of revealing the functioning of the brain, but that situation began to change once recording electrodes left the surface and entered the brain tissue (see Chapter 8). Whenever a neuron fires an action potential it broadcasts an electrical signal into the surrounding volume conductor. That signal is larger closer to the neuron, and electrodes inserted into the brain can detect the activity of nearby neurons. A relatively large electrode can detect the summed electrical activity of a relatively large number of neurons, but a fine electrode insulated except for the tip can detect action potentials of individual neurons. Studies of the electrical activity of individual neurons have revealed significant information about how the brain works (see Chapter 8), and the technology of implanted electrode arrays is being developed, which promises to transform the lives of paralyzed people and many others with damage to their nervous systems resulting from disease or injury (see Chapter 10).

THE FRINGE

My book is focused on explaining generally accepted mechanisms, but science always has been and remains filled with ideas that are on the fringe. Some look like and are crackpot ideas, but the history of science is full of examples where fringe science eventually graduated to mainstream science. My favorite example is graduation of the crackpot idea of Alfred Wegner (1880–1930) that the Americas were once joined with Europe and Africa and there was no Atlantic Ocean into the modern concepts of sea-floor spreading and continental drift.

There is a huge fringe in the biological sciences extending from supernatural and psychic phenomena (parapsychology) to the more down to earth: for example, from experiments into extrasensory perception (ESP) and psychokinesis to experiments into the possibility that electric fields produced in the body control nerve growth. ESP is an interesting case for scientific study. The concept of extrasensory perception is a bit of a contradiction since science seeks mechanisms, and *extrasensory* implies perception without a sensory mechanism. Of course, the idea is that the mechanism of ESP is outside of the known modalities of sensation, which many would believe makes it supernatural. Thus, an experiment unequivocally demonstrating the reality of ESP would imply to some the existence of a supernatural world. But science cannot prove the negative in this kind of circumstance, so it is always possible that basic research could eventually find a mechanism for ESP that is within the realm of scientific explanation. The subject area of parapsychology includes ESP and psychokinesis, along with investigations into the survival of consciousness after death, near-death experiences, out-of-body experiences, and even reincarnation, so the association of parapsychology with the supernatural is strong. Nonetheless, parapsychology has straddled the fringe of academic research for many years with a presence at some major universities, for example, Duke University, including the establishment of research chairs. The Parapsychological Association, under the direction of none other than the anthropologist Margaret

Mead, became affiliated in 1969 with the American Association for the Advancement of Science, none other than the publisher of the prestigious journal *Science* (see Wheeler 1979).

The two factors that determine the fringiness of an area of "scientific" inquiry are the quality of the data and the availability of possible mechanisms that can be investigated. The 1960s, 1970s, and 1980s witnessed many debates about the quality of the data supporting parapsychological phenomena, and the possible mechanisms were firmly rooted in psychic outer space, so without unequivocal proof, interest eventually died away. One can ask why interest in the supernatural has persisted for so long in our supposedly scientific world. There are undoubtedly many reasons; the possibility of immortality is likely the most compelling. The human propensity to believe in the supernatural is treated eloquently in Carl Sagan's book *The Demon-Haunted World* (1996). The possibilities of telepathy and mind reading have implications that are relevant to defense and the military, so even the remotest possibility has occasionally attracted research funding. There are also likely to be medical, psychological, and social implications; the possibilities of witchcraft and casting of spells comes to mind.

So what does all this tell us about science, which is after all a human activity? Before the scientific revolution, false explanations such as the concept of animal spirits lasted for millennia. It is important to realize that after the advent of the scientific method false explanations can still last for decades. A second key realization is that nonscientific explanations die because they don't lead to fruitful research and scientists lose interest and funding, not necessarily because they are proven wrong. If solid scientific evidence for some parapsychological phenomenon is found, parapsychology could come back and eventually even become part of the mainstream, but because of its history few scientists are researching parapsychological phenomena anymore.

There are other fringe areas of scientific inquiry that are a lot less fantastic. It is not impossible that the electric fields projected throughout the body by electrically excitable cells could themselves have a function. Are brain waves, for example, actually signals that carry information from one brain region to another? This seems unlikely, since small external currents do not easily cross cell membranes, which have high electrical resistance relative to the extracellular fluid, and therefore broadcast signals have little or no effect on the membrane potential. There do not seem to be any cells in mammals specialized for the sole purpose of broadcasting amplified currents. While it is impossible to absolutely rule out the possibility that heart waves, brain waves, muscle waves, or other waves have some physiological functions, the mainstream belief is that the function lies with the cellular processes that produce the waves, and the waves are just low-level epiphenomena without any function of their own.

Do some cells have a means of detecting electric fields? The answer is that in principle all cells can detect relatively steady electric fields. Recall from Chapter 2 that the membrane is a fluid mosaic with proteins and lipid molecules floating in the plane of the membrane. Proteins have specific functions, and unless a membrane protein molecule is otherwise anchored inside the cell or to the extracellular matrix outside the cell, it is free to float over the surface of the cell. Proteins are electrically charged at a magnitude and sign that depends upon the number of negatively and positively charged amino acids in the structure of the protein. Proteins that are not anchored and have a sufficient excess of negative or positive charge will move along the membrane if the cell is exposed to a steady electric field. This means that an electric field could cause a given membrane protein to accumulate on one part of the cell and therefore cause the function of that protein to localize at that part of the cell.

Electric fields are also produced across the membranes of all cells, and there are examples where cells arranged in sheets such as epithelial cells can produce a significant steady electric field. A prime example is the skin of a frog. Frog-skin epithelial cells have ion pumps in their membranes that transport ions across the skin between the inside of the frog and the surrounding pond water. This is a normal function, which helps maintain a higher concentration of ions in the blood and extracellular fluid of the frog than in the surrounding water. As a result, a steady voltage develops across the frog skin on the order of tens of millivolts, with the inside of the frog negative relative to the pond water. If the frog suffers a penetrating injury to the skin, the voltage across the frog skin is short-circuited, and a current loop develops with current entering the frog across the intact skin and departing through the wound. It is not clear if this current has any function. Such currents occur normally in the frog embryo during development where a steady current flows out of a hole in the early embryo called the blastopore. The production of steady electric fields is a widespread phenomenon in organisms and may have biological importance.

There was a flurry of interest in the possible use of steady electric fields to promote regeneration of amputated limbs (see the review by Borgens et al. 1989). In frogs, which normally do not regenerate limbs, amputation of a forelimb results in the growth within about 6–10 months of a narrow, pointed extension from the stump with a core of cartilage. It doesn't much resemble a normal limb. If an electric field is applied that produces current flow out of the stump during this period, the extension is fatter along its entire length, is club-like rather than pointed, and contains some regenerated axons. But it is clearly not anywhere close to a regenerated limb in all its complexity. It was also reported that if the fingertip of a child is accidentally amputated—just the tip beyond the last joint—regeneration of a completely normal fingertip can ensue, provided that the usual procedure of suturing the stump closed is not done so that the wound is left open. One theory is that electric current leaving the cut end of the fingertip promotes regeneration, which is blocked when the end is sutured closed. Current has been measured leaving the cut end of a finger. If you are wondering where it goes, it must complete the circuit by spreading out in the moisture on the surface of the skin and then diffusely reentering the body. The published picture of the stump (reproduced in Borgens et al. 1989) after the fingertip amputation shows the view of the palm side of the finger so the extent of nail bed remaining is not visible. The picture of that same finger after 12 weeks, showing the regenerated fingertip, provides an end-on view of the fingertip with the nail bed partially in shadow. In contrast, the photograph of a finger that was sewn over after a similar accidental amputation clearly shows a disfigured fingertip with no nail bed after four months. One would like to have seen comparable and well-lighted before-and-after views of both fingertips showing the status of the fingernails. The evidence as presented seems to me consistent with the possibility that the apparent difference in outcomes might have resulted just from disfigurement caused by suturing in one case and not the other. While regeneration of human limbs, even fingertips, seems currently out of reach, electric field stimulation has numerous effects on cells that are conducive to growth and repair and, for example, is used clinically to aid in the healing of broken bones (Ciombor and Aaron 2005).

Much interest was generated by a report that artificially imposed electric fields can affect the direction of axon growth from cultured neurons. Lionel Jaffe and Mu-Ming Poo (1979) placed chunks of developing dorsal root ganglia from chick embryos in a tissue culture dish with suitable liquid medium that kept the sensory neurons alive and promoted the growth of axons. The axons grew out on the floor of the dish in all directions. Jaffe and Poo placed electrodes in the culture medium on each side of the dish, applied an electric field of 70–140 mV/mm, and observed the effect on axon growth. The axons grew at an average rate of 26 micrometers (μm) per hour from the side of the chunk of tissue facing the negative electrode, compared to 10 μm per hour on the side facing the positive electrode. So it appeared that axon growth was a little more than two and a half times faster in the direction of positive current flow between the two electrodes.

Any time an experimental treatment is shown to have an effect on axon growth, the first thought is to figure out how to use it to promote regeneration in people with spinal cord injuries. After numerous animal studies, application of an electric field in order to promote axon regeneration after spinal injury in humans was the subject of a phase 1 clinical trial, the results of which have been published (Shapiro et al. 2005). Ten spinal-injured patients received 15 weeks of electrical stimulation at a very low voltage delivered between electrodes surgically implanted at the vertebrae just above and just below the injury site. The polarity of the field was reversed every 15 minutes during the entire 15-week period. The rationale for alternating the polarity was that it would promote axon growth in both directions. Some patients displayed a degree of improvement on neurological assessment measures one year after treatment. I am not aware of any further published reports of clinical trials on this subject during the ensuing decade. For now the evidence suggests that electric fields are not likely to be useful in promoting spinal cord regeneration (see Robinson and Cormie 2007).

SPECULATION

I do wonder, though, if electric field effects have been thoroughly explored. Experiments that I published in the 1980s occasionally still stimulate my thoughts about the subject (Campenot 1984, 1986). I had developed a chambered system for cultured neurons where sympathetic neurons from newborn rats are placed in a central compartment, and their axons grow to the left and right, cross under silicone grease barriers, and continue to grow within left and right distal compartments. For a number of reasons I investigated the effect of the potassium concentration in the culture medium on axon growth. Normally the potassium concentration is low, and when it is raised, the axonal membrane is depolarized. I observed that elevating the potassium concentration in a distal compartment reduced the number of axons that entered, and when applied to axons already within a distal compartment elevated potassium caused growth to stop and the axons to retract and degenerate. Distal axons do not like elevated potassium, so what's the big deal? Well, none of this happened if the center compartment containing the cell bodies and proximal segments of the axons was also supplied with high potassium. Nor was there any degeneration if the cell bodies and proximal axons were supplied with high potassium and the distal axons were supplied with normal potassium. The only configuration that the neurons did not like was elevated potassium only on the distal axons. Elevating potassium depolarizes the distal axons. When the cell bodies and proximal axons are not given high potassium, they are not depolarized, so a longitudinal electric field develops inside the axons with the distal axons positive relative to the proximal axons and cell bodies. This internal electric field would drive internal current toward the cell bodies, which interestingly is the opposite direction of the internal current produced by the externally applied electric fields that have been shown to promote axon growth.

Could variations in the potassium concentration in the extracellular fluid set up currents along axons that affect their growth and survival? That is not known, but it seems possible to me. Here is one speculative scenario: Neurons release potassium into the surrounding extracellular space every time they fire an action potential, presumably producing a fairly uniform increase in the extracellular potassium to which their entire surfaces are exposed. Now consider a presynaptic neuron with little or no electrical activity whose axon terminals converge onto a postsynaptic neuron that receives synapses from a number of more active neurons. Every time the activity of the more active neurons causes the postsynaptic neuron to fire, the terminals of the less active neuron might receive a dose of extracellular potassium directly on their axon terminals. If the terminals respond to elevated potassium anything like the distal axons respond to high potassium in culture, then repeated doses of potassium might cause terminals of the less active presynaptic neuron to retract and degenerate. The presynaptic terminals of the more active neurons might be protected from degeneration because their activity is coincident with the firing of action potentials by the postsynaptic neuron and would be expected to raise the extracellular potassium along the entire length of their axons, not just at the terminals. It is a well-established principle that active synapses can displace inactive synapses from postsynaptic neurons (see Chapter 8), so it is at least conceivable that electric fields set up by potassium released during activity could play such a role in circuit formation and maintenance.

This is all speculation, but speculation is an important part of scientific research. Interest has moved away from the study of the effects of steady electric fields on phenomena such as nerve growth, but that does not mean that there are no important effects yet to be discovered. Experiments are hard to do, they are expensive, and obtaining negative results leads neither to grant renewals nor to progress of the careers of the experimenters, of whom many are graduate students or postdoctoral fellows. Research areas that have been explored and have not so far yielded important information attract few takers.

WHAT SCIENTISTS KNOW

The topics covered in this chapter illustrate the large variations that exist in human scientific knowledge and something about why such variations exist. The mechanisms involved in generating the ECG are very well known because of its relative simplicity and its medical relevance. The mechanisms by which the torpedo ray and electric eel generate strong shocks are also very well known. The reasons are somewhat different. The torpedo ray was a focus of investigations from the beginning of inquiries into animal electricity because of its dramatic and mysterious nature—we have to know how torpedo rays do this or we really do not understand the fundamentals of how nature works. Also, as it turned out, investigations of the torpedo ray contributed to basic knowledge about how synaptic transmission works by providing a source for the purification of acetylcholine receptor channel proteins and other molecules involved in neuromuscular transmission. The torpedo ray was one of those animals where nature is revealed; most animals have done a much better job of hiding the electric nature of their functioning. When we move away from the ECG and the torpedo ray and electric eel, knowledge gets murky fast. Investigations into the "meaning" of the EEG never panned out. It is not at all clear whether electric fields generated in the body have any direct function—they look like epiphenomena. The extent of the use of electric fields by aquatic animals is unclear, and the precise mechanisms of electroreception are not very well known because a large range of animals are involved, many of which are hard to obtain, and even if the animals are obtained, the mechanisms are hard and time-consuming to study. Without the promise of direct medical relevance, it is difficult to finance the tremendous effort that is required.

So it is important to realize something about scientific inquiry that is so easy to forget in our world where so much is known. We biological and medical scientists tend to believe that we know most of the basic principles operating in animals, so variation arises because of differences in which mechanisms are emphasized and how they interrelate. Maybe true; maybe not; but it is clear that we are ignorant. Knowledge can be counted and weighed in many ways, and knowledge of basic principles undoubtedly should count more than knowledge of details. But regardless of that, there are more details than basic principles, and every now and then it happens that something new and important breaks through. Human impact on the earth is so great that we need to keep our eyes and our inquiring minds open so as not to miss something that could improve our well-being and survival and that of the planet and its wonderful diversity of inhabitants.

10

THE BIONIC CENTURY

A major problem with recovering from injuries to the nervous system arises from the severing of long axons. Consider the circuit that allows me to move my fingers to type on my computer keyboard. Somewhere in my brain circuitry I generate the intention to type the letter *a* using the little finger of my left hand. The circuit of motor neurons connected to the muscles in my forearm that causes my little finger to press the *a* key begins with a population of motor neurons in the motor cortex, whose axons extend about 20 centimeters (cm) down the spinal cord and synapse on a population of spinal motor neurons. The axons of the spinal motor neurons extend about 50 cm from the cell bodies out into the peripheral nerve to the muscles in my forearm that pull on the tendons to move my fingers. If you were to interrupt this pathway by cutting off your little finger in a table-saw accident, compose yourself, stop the bleeding, pick up your little finger, put it in a Ziploc, put it on ice, and get to the emergency room as soon as possible. Since the nerves and muscles that move your fingers are located in your forearm, not in the fingers, restoring movement is strictly a mechanical problem. Provided that the finger is not too mangled, the chances are good that a surgeon can sew your finger back on, reattaching the tendons and blood vessels, and you will be typing on your computer again in the not-too-distant future.

Injuries higher up the arm are more complicated if the nerve is severed. Severing the ulnar nerve above the elbow severs the axons of the spinal motor neurons, and the cut-off segments of the axons soon die, denervating the muscles in the forearm that move the fourth finger and the little finger. The spinal motor neurons do not die, and the ulnar nerve below the cut, being made of many cells and connective tissue, remains but is devoid of motor axons. If a surgeon reconnects the ulnar nerve, the motor axons can begin to regenerate across the cut and grow back to the muscles in the forearm that control the fourth finger and little finger. This can take considerable time, since axons grow slowly, about 10 cm in 100 days, so the closer the cut end is to the muscle the better the outcome. By carefully sewing the ends of the ulnar nerve together in the original orientation, reasonable function may be restored—enough to type again, but maybe not to enough to preserve one's career as a concert pianist.

THE PROBLEM OF NERVE REGENERATION

I remember learning long before I became a neuroscientist that the reason an injury to the spinal cord often results in permanent disability is that neurons in the central nervous system cannot be replaced—you have only the number you are born with—so if they are injured or killed, that's it. We now know that neurons are produced from stem cells in the brain and spinal cord throughout life, so this is not necessarily the reason spinal cord injuries cause permanent disability. The problem is damage to the long axons that extend from cortical motor neurons down the spinal cord. The cell bodies of spinal motor neurons in the lumbar region of the spinal cord whose axons innervate the leg musculature are about a third of a meter below the level of the neck. Thus, if the neck is broken, severing the axons of the cortical motor neurons, their axons would have to grow a distance of about a third of a meter to reach the vicinity of the cell bodies of the spinal motor neurons that they innervated before the injury. Once there, the regenerated axons would have to branch appropriately and form synapses on the motor neurons they originally innervated. This seems like a tall order, but it seems especially difficult because never before have the axons of the cortical motor neurons grown such a long distance to reach their target spinal motor neurons; the distances within the embryo were much shorter when these connections were originally established. If one considers axon growth from the time axons first appear during embryonic development and the time at which humans reach full body size, only a tiny fraction of axon growth occurs when the axons are growing out and finding their innervation targets during development. Rather, it is rarely realized that most of the axonal elongation that occurs in the body is produced as the already connected nervous system enlarges in size to keep pace with the growth of the body. It is well established that axons elongate at rates that are maximally about 1 millimeter (mm) per day. At that rate, axon growth spanning the third of a meter between the level of the neck and the level of the cell bodies of the lumbar motor neurons would require at minimum almost a year without taking into account obstacles such as scar tissue at the injury site. A year is longer than the entire gestation period of the embryo during which the entire nervous system develops. A lot can happen in a year that may not be conducive to restoring the functions lost because of the injury. Nonetheless, promoting regeneration of axons in the spinal cord has been a primary objective of researchers attempting to discover ways to restore function after spinal injury. However, severed axons in the central nervous system don't seem able to grow much at all.

In contrast, axons outside the central nervous system, such as the axons of the spinal motor neurons that innervate skeletal muscles, do regenerate after being severed. This is true for humans and rats and likely true for all mammals. Why the difference? An insight came to me while thinking about this problem in evolutionary terms. Suppose you are an animal, for example, a wildebeest that is subject to predation by lions and hyenas out on the African savanna. Now suppose you are a very young wildebeest still watched over by your mother, and you were injured somehow. Say a hyena took a bite out of you and you managed to survive because your mother succeeded in chasing it off. If the bite damaged your spinal cord, you have had it. First of all, a bite that deep will probably cause you to bleed out, but if you did survive until morning you would probably be breakfast for a pride of lions against which your mother could offer no protection; even if the spinal axons could regenerate, there is no way you could survive long enough for their regeneration to restore your mobility.

Suppose the hyena just bit your leg high up on your thigh, severing the sciatic nerve that innervates much of the leg musculature. Would the outcome be any different? The answer is probably not. If the major artery supplying the leg with blood was severed, you would bleed out, and, if not, the loss of mobility resulting from a disabled leg would turn you into predator food in short order. So, even though the axons in the sciatic nerve can regenerate after being severed, there is no way you could survive during the months required to restore the function of your leg. Thus, it is hard to see how even peripheral nerve regeneration could have evolved, since it is unclear how it could confer a survival advantage to animals without access to a health care system to assist their recovery.

Now suppose that the bite of the hyena was less severe, piercing the skin and lacerating one of the leg muscles that is innervated by motor axons from the sciatic nerve. Suppose you were lucky and there was no excessive bleeding or infection. The laceration will have severed some of the muscle fibers and also possibly severed some of the axon branches that innervate the muscle. The severed muscle fibers will be replaced by new ones produced by stem cells that reside within the muscle, and these new muscle fibers will need to be innervated by branches of the motor axons or they will not function. Therefore, recovery from a muscle laceration likely requires the regeneration of axon branches to re-innervate muscle fibers that had been denervated by the injury and also to innervate new muscle fibers that replaced those that were damaged by the injury. Thus, my theory of why peripheral axons regenerate and central nervous system axons do not is that peripheral axon regeneration is necessary for full recovery from relatively minor injuries. When the teeth of the hyena lacerated your leg muscle, the function of the muscle was reduced because the function of some of the muscle fibers was temporarily lost, but the remaining undamaged muscle fibers that were still innervated by intact branches of sciatic nerve axons were still operational, so the muscle still functioned, albeit at a reduced strength. You could still walk and even run. Since only relatively short branches of the sciatic axons were cut, regeneration leading to restoration of full function could occur in a reasonable amount of time. If you could lay low and with the help of your mother avoid the attention of hyenas and lions, you could recover full function and go on to pass your genes to your own offspring. Now we can see an evolutionary advantage to peripheral nerve regeneration. Possibly, then, the capacity for regeneration of longer lengths of peripheral nerve just came along for the ride with the capacity to regenerate nerve endings within muscle.

Now let's return to the sphere of experimental science. In 1981 an experiment by Sam David and Albert Aguayo at McGill University in Montreal set the stage for experiments into spinal cord regeneration for over 30 years (David and Aguayo 1981). The question was a simple one: Is it some intrinsic property of the neurons in the central nervous system that prevents them from regenerating axons after injury, or is it their environment? David and Aguayo approached this question by asking if axons of central nervous system neurons can regenerate within a peripheral nerve. They performed a surgical rearrangement by grafting a 35-mm length of sciatic nerve taken from the leg of a rat onto the rat's spinal cord, like a teacup handle, forming a bridge connecting the brain stem at the base of the brain to the spinal cord in the cervical (neck) level or upper thoracic (chest) level. They gained access to the brain stem and spinal cord by removing some bone, and used a glass rod as a plunger to stuff the ends of the sciatic nerve through, securing them in place with sutures. This procedure caused some local damage to the brain stem and spinal cord at the sites of insertion. Importantly, the sciatic nerve used for the bridge initially contained only severed axon segments since the nerve cell bodies that give rise to the sciatic axons reside in the spinal cord in the case of motor neurons or in peripheral ganglia in the case of sensory axons. All those severed axons in the sciatic nerve bridge degenerated, leaving mostly connective tissue, the cells that produce connective tissue, and Schwann cells, which cover the surfaces of all peripheral axons and myelinate some of them.

After 22–30 weeks the sciatic nerve bridge was cut near the middle, and the tracer enzyme HRP (see Chapter 7) was applied to both the cut end connected to the brain stem and the cut end connected to the spinal cord. After allowing time for the uptake and transport of HRP, the parts of the brain stem and spinal cord with the sciatic nerve bridge attached were removed and examined by light and electron microscopy. The results showed that axons had grown from cell bodies of neurons in the brain stem and spinal cord and extended long distances within the sciatic nerve bridge. These newly grown axons originated from neurons whose axons normally remain entirely within the central nervous system.

These results indicated that axons from neurons in the brain stem and spinal cord of the adult rat have the ability to grow long distances within the environment provided within the sciatic nerve bridge. It was not determined whether the axons in the bridges arose by regeneration from axons damaged during the grafting procedure or from sprouting from undamaged neurons. Either way, the environment provided by the Schwann cells and other cells and connective tissue in the sciatic nerve bridge was conducive to axon growth, unlike the environment within the brain and spinal cord. It is not clear whether the axons of the neurons that innervated the sciatic nerve bridge had to grow a short distance within the brain stem or spinal cord to reach the environment of the bridge since the growth may have originated from cut ends of axons at the entrance to the bridge. Also some Schwann cells from the bridge had migrated into the brain stem and spinal cord, so the environment in the brain stem and spinal cord within a few millimeters of the bridge insertion sites may have been altered and become more conducive to axon growth.

The next questions were, of course: What makes the brain and spinal cord nonpermissive for axon growth, and can this be changed so as to allow axon regeneration within injured spinal cords resulting in functional recovery in people with spinal cord injuries? There is reason to believe that prevention of axon growth is actually as important a mechanism during the embryonic and postnatal development of the central nervous system as the promotion of axon growth. The complicated wiring of the developing central nervous system is established in a defined sequence where axon growth in some tracts is established early and then ceases when the pathways are completed and then the growth of axons in other tracts can begin. Thus, it seemed that there may be signals that appear during development that turn off axon growth once a tract is completed. These signals may persist in the adult central nervous system, serving to prevent aberrant, uncontrolled growth once the major circuits are complete.

The results of David and Aguayo prompted many investigators to get into the act; funding became available partly because of the implications for curing paralysis caused by spinal injury. It was relatively quickly discovered that several membrane proteins that inhibit axon growth were produced by oligodendrocytes, which are the cells that give rise to myelin in the brain and spinal cord. These molecules are exposed on the cell surface and cause axon growth to cease upon contact. This effect can be neutralized by the introduction into the central nervous system of antibodies to these axon-growth-inhibitory molecules. Many such experiments have been conducted. Neutralization of growth-inhibitory molecules can produce very limited axon growth across a damaged section of rat spinal cord where the original axons have been severed, but the growth beyond the injury remains very limited in extent—nothing like the extensive axon growth that occurred in the sciatic nerve bridges.

So far it seems that regeneration producing something similar to the original pathways and connections that were present before injury may not be possible. Now most neuroscientists are thinking more about generating the growth of new pathways consisting of shorter axons to restore function. The spinal cord is not just a cable of axons; it contains lots of neurons and a lot of internal circuitry. For example, subsystems of circuitry within the spinal cord control aspects of your locomotion, such as alternating stepping when you walk. In effect, your spinal cord "knows" how to walk; your brain tells it when and where and modifies the fine points of how you take each step. The motor axons from the cortical motor neurons each make a multitude of connections including, but by no means exclusively, on the spinal motor neurons that carry the final signals to the leg muscles. The terminals of the cortical motor axons also synapse on many interneurons whose connections are entirely in the spinal cord and which participate in processing this information before it is delivered to the muscles by the spinal motor neurons. Interestingly, the decisions made within the spinal cord are final, and once the information has reached the spinal motor neurons and determined their pattern of firing, all aspects of the movement have already been determined. The sensory information required for walking, such as the touch of your feet to the surface, proprioceptive information about the positions of your legs and feet, and even visual information such as when you watch your feet while going down the stairs, is all delivered to the spinal cord and incorporated into the final excitation of the motor neurons.

As you might imagine, it is hard to investigate the subsystems in the human spinal cord that participate in walking because humans cannot be subjected to the invasive experimental procedures that are used on animals. Nonetheless, the central idea is the hope that limited growth of axons of cortical motor neurons and/or spinal interneurons across a spinal cord injury site may produce synapses onto neurons that are part of the subsystems that control movement of the paralyzed limbs and restore some function, even if the axons cannot grow all the way to re-innervate the spinal motor neurons to which they originally connected. An underlying presupposition of this hope is that even though the new connections are not the original ones and may not initially restore the lost function, once connected in what are essentially new circuits, and some willful movement is restored, that the new circuits will be subject to learning, which can ultimately result in recovery of some meaningful function.

This may sound like a limited goal that is unlikely to restore walking, but walking isn't everything. Most spinal cord injuries are caused by broken necks and result in loss of some or all function of the arms and hands. Spinal cord injuries are usually not complete transections, which sever all the axons at the injury site, and what axons and how many are spared determine the function that remains. After a little thought it will be obvious that partial function of the arms and hands is a lot more useful than partial function of the legs and feet. Legs and feet must function at a high level for walking, but even a relatively low level of function of arms and hands can allow a partially paralyzed person to independently take a drink, eat, and operate electrical controls, which makes a big difference in independence and overall quality of life. Thus, if some limited regeneration led to enhancement of arm and hand function, then that would be a major breakthrough. Also, even partial enhancement of bladder and bowel control could make a big difference in quality of life.

I believe that the view of many people that spinal cord injury should ultimately be completely reparable arises partly because spinal cord injury is not visible on the outside of the body unlike, for example, amputation. It is hard to imagine amputees regrowing legs and walking, and there is no serious research being conducted of which I am aware aimed at producing this outcome. Severe spinal cord injuries are hidden inside the body so it is less difficult to imagine a paraplegic person getting up out of the wheelchair and walking.

MIND-READING PROSTHETICS

Imagine that a rhesus monkey with a computer wired to his skull and his right forearm is sitting in a chair performing a task. There is a ball on a tray in front of him. He picks up the ball by grasping it with his right hand, raises his arm, and releases the ball into a vertically oriented tube that leads back to the tray. The ball rolls out onto the tray. During his training on the task he has received many food rewards, so he immediately picks up the ball again and repeats the task. Then one of the experimenters injects a local anesthetic into the monkey's right forearm, which paralyzes the monkey's right hand. He can still move his arm at the shoulder and elbow joints, but the hand is paralyzed, so he can no longer perform the task. Then one of the experimenters turns on the computer equipment wired between the monkey's skull and forearm, and the use of the right hand is restored. The monkey grasps the ball and places it in the tube. The ball rolls out onto the tray, and the monkey picks it up and repeats the task. When the computer equipment is turned off, the hand immediately becomes paralyzed again. Thus, it appears that the computer connected between his skull and the forearm had restored the ability of the monkey to grasp the ball.

In the actual experiment wires connected to recording electrodes in the motor cortex detected action potentials in neurons that had been activated by the monkey's intention to grasp the ball (Ethier et al. 2012). The computer responded by sending a train of electrical pulses to electrodes inserted into the muscles in the monkey's forearm. The pulses depolarized the membranes of the muscle fibers to threshold, producing action potentials, which caused the muscle fibers to contract, pulling on the tendons that flexed the monkey's fingers. Thus, the computerized circuit between the monkey's motor cortex and forearm muscles bypassed the spinal cord and peripheral nerves and turned the monkey's intention to grasp the ball into the act.

The mystery of how this works begins with the question of how electrodes in the motor cortex detected the monkey's intention to flex his fingers to grasp the ball. Modern knowledge of the organization of the motor cortex began with the functional maps of the motor and sensory cortices produced by neurosurgeon Wilder Penfield (1891–1976) of the Montreal Neurological Institute. He electrically stimulated the surface of the primary motor cortex while he was performing surgery on patients suffering from epilepsy and recorded the body movements that were produced as the stimulating electrode was moved from one location to the next. Patients are not given general anesthetic during brain surgery since there are no pain receptors in the brain, so he was also able to map the cortex involved in the sense of touch by asking patients where on the body they felt a touch when locations in the somatosensory cortex were electrically stimulated. He discovered that the motor cortex is organized so that motor neurons producing movement in adjacent parts of the body are adjacent to each other in the motor cortex. This means that the anatomical relationships of the parts of the body would look something like a person when mapped out on the cortex—a very distorted person as it turns out since regions capable of precise movements like the face, hands, and fingers occupy proportionately more cortex than regions where control is less precise, such as the upper arms. Penfield also found that the surface of the body was mapped out on the surface of the primary somatosensory cortex.

Your every movement requires that some of the neurons in the primary motor cortex of your brain become activated. Different populations of neurons are activated to produce different movements. Even the simplest movements, such as pressing down with your little finger to type the letter *a,* are commanded by the activity of hundreds of thousands to millions of neurons in the primary motor cortex. It is not possible to monitor the individual activity of all the neurons that are involved, so the exact signal for pressing the little finger down on the *a* key cannot be determined. Moreover, it is not enough to just specify that the little finger is to be pressed down since that will not accomplish the task if the finger is not located over the *a* key—the hand needs to be in the correct position. Sensory feedback also plays a role since the feel of the key as it is depressed is also involved in ensuring the success of the task. Expand the task to touch-typing an entire sentence, and the number of neurons in the cortex that are involved likely expands into the billions.

However, the vote of every neuron need not be counted in order to know that a command has been issued to press down the little finger. There is debate in the field about how many cortical neurons must be monitored for a high degree of accuracy of the predicted movement, but it seems likely that a few hundred will suffice. A major goal of the field of neuroengineering is to develop computer algorithms that can read the hand and arm movements being commanded by the brain in the activity recorded from cortical neurons and use this information to command the same movements in a robotic arm or prosthetic device or even in the paralyzed hand and arm themselves.

The researchers that performed the rhesus monkey experiment described above had set out to detect the activity patterns of individual neurons whose firing is associated with the intention to grasp a ball and transform the recorded intention into a pattern of direct muscle stimulation that produced grasping by the hand. Since only the brain and the muscles are involved, this experiment serves as a model for electronically bypassing the spinal cord and peripheral nerves in a human with a spinal cord injury. In order to detect the monkey's intention, they used an electrode array consisting of a square grid about 4 mm on a side, containing 100 very fine electrodes inserted to a depth of about 1.5 mm into the motor cortex. Each electrode in the array detects action potentials from cortical neurons nearby, similar to the tungsten electrodes used by Hubel and Wiesel in their studies of the visual cortex (Chapter 8). The closer a neuron is to the electrode, the larger the amplitude of the detected action potential, and most neurons detected by the array broadcast their action potentials to more than one electrode. The computer algorithm used picks out the pattern of detection at multiple electrodes produced by the firing of each individual neuron. Using this approach, firing patterns of approximately 100 neurons were simultaneously detected and analyzed.

Inserting an electrode array into the cortex of a human is an invasive procedure. The electrical activity of the cortex can also be detected by electrodes placed on the surface of the head that record the electroencephalogram (EEG), which is completely noninvasive, or by recording from electrodes placed under the skull on the surface of the brain, called an electrocorticogram, which is far less invasive than inserting electrodes into the cortex. In both cases the recording electrodes are too far away from the source of the signals to detect action potentials of individual neurons. The EEG and electrocorticogram waves are produced by the combined synaptic potentials, called field potentials, arising in populations of neurons. The principle advantage of the electrocorticogram is that the signal is not attenuated by passage through the skull as is the case for the EEG. While changes in the pattern of the EEG are associated with changes in behavior, many researchers in the field view it as unlikely that the association of changes in the EEG or electrocorticogram with any precise intention such as grasping the fingers is close enough to use the EEG to command prosthetic circuitry to produce the intended movements.

After an electrode array is in place and monitoring the activity of neurons in the motor cortex, the next task is reading what the activity pattern means. In the rhesus monkey experiment this was relatively easy since the monkey did not have a paralyzing spinal cord injury, and only one intended movement, grasping, needed to be identified. The monkey was trained to repeatedly pick up a ball and place it into the tube. During the monkey's training the activity of the cortical neurons was recorded. The repeated activity pattern that was associated with the task was taken as reflecting the monkey's intention to grasp the ball.

Because paralyzed humans cannot perform the intended movements, the brain activity associated with the subject's intention must be identified in another way, such as monitoring the activity while the subject imagines making the movement. How the subject does it, though, is immaterial, as long as the subject can reproduce the same pattern of electrical activity when he or she intends to make the desired movement. For example, paralyzed subjects can learn to control a robotic arm by modifying the electrical activity of the monitored neurons. Whether the electrical activity of the neurons in any way represented the original "intention" of the cortex to move the subject's real arm or an entirely learned intention to perform an entirely new task is perhaps irrelevant. Instead, it may be best to consider the motor cortex as a population of neurons under conscious control that because of its plasticity can be tapped to control a wide variety of new functions.

The next step in the monkey experiment was to link intention and movement together through the computer running an algorithm that transforms the monkey's intention into the pattern of pulses to the muscles that produce grasping by the fingers. With the computer algorithm in place, the investigators temporarily disconnected the monkey's hand from its brain by injecting local anesthetic to block transmission in the median and ulnar nerves that contain the motor axons that innervate the muscles in the forearm that move the fingers. With the hand paralyzed, the monkey was still able to make the intended grasping movement using the external electrical connection between its brain and muscles provided by the experimenters.

Progress must be made on multiple fronts before this kind of approach can be routinely used in a clinical setting with humans (Lebedev and Nicolelis 2006). Electrodes damage neurons when they are inserted into brain tissue, but there are lots of neurons, so one can afford to lose a few. But can electrodes be implanted in the brain and left there for years, continuing to record the activity of enough neurons to be useful? Electrodes cause acute immune reactions within brain tissue, called the foreign body response, and long-term glial scarring (Polikov, Tresco, and Reichert 2005). Glial cells are the most abundant cells in the brain. Since an electrode needs to be within about 0.1 mm of the cell body of a cortical neuron to record an action potential, glial cells gathering around an electrode and encapsulating it with a glial scar can get in the way. However, electrodes have been implanted in patients for a procedure called deep brain stimulation, which has been used successfully for the treatment of symptoms in Parkinson's disease and is being used increasingly for treatment of other disorders of brain circuitry (Lozano and Lipsman 2013). In the treatment of Parkinson's disease, electrodes are implanted into subcortical regions involved in the control of movement, and long-term electrical stimulation alleviates symptoms such as tremor. In a study involving the examination of brain tissue after death, neuronal cell bodies were located near the electrode tracks, and there was only a mild response of the glial cells that was not progressive, even though the patients had received continuous deep brain stimulation for up to 70 months (Haberler et al. 2000). Recording electrodes must be in closer proximity to the neurons than stimulating electrodes, which apply higher voltages than are produced by the neurons, but even so these results are encouraging. If glial scar formation is a problem in some circumstances, one approach might be to preempt glial scar formation and get the neurons to snuggle up to the electrodes by providing a coating of bioactive molecules such as neurotrophic factors (see Chapter 8).

In the monkey experiment above, the intention read in the brain was fairly simple: contract the flexor muscles of the five fingers. The electrodes in the muscles were set up to do just that, so the computer algorithm was fairly simple too. The actual placement of the ball in the tube was accomplished by moving the arm with the shoulder and upper arm muscles, which were not paralyzed by the local anesthetic. Restoring useful movements of an entire paralyzed arm is much more complex, since it would require multiple electrode arrays that deliver stimulation that could discriminate the intentions to perform coordinated arm and hand movements. In a prosthetic brain-to-muscle circuit fully developed for human use, the electrode arrays would be connected to a miniature transmitter on the surface of the brain that broadcasts the signals wirelessly to a miniature computer embedded near the target muscles. The computer would be connected with wires to stimulating electrodes in the muscles. Thus, there would be no chronic incisions into the body or the central nervous system.

There is a history beginning in the 1950s of successful muscle stimulation in a clinical setting in the form of artificial cardiac pacemakers. Pacemakers generally consist of an implanted, battery-operated control module that connects to electrodes attached to the heart muscle through wire leads. The control module regulates the delivery of stimulation through algorithms that, for example, only supply stimulation when the heart rate falls below a selected level. Pacemakers have now been developed that enclose all the functions in a single cylindrical module about 5 mm in diameter and 20 mm long, designed to be delivered to a site on the heart through an intravenous catheter (Cheng and Tereshchenko 2011). Muscle stimulation such as that used in the experiment described above requires external control. One can imagine the future development of tiny stimulation modules with self-contained power supplies, implanted in each of the muscles involved in complex hand movements, and responding wirelessly to the coordinated commands of a control module.

Neural prosthetic devices easily become the subject of futuristic imaginings, and it is hard to predict what the future might bring. Clearly there is a potential role for restoring hand movements—the more precisely controlled, the better—but restoration of even crude hand movements could be of tremendous benefit. Hand movements that would allow a quadriplegic individual to eat and drink independently, drive an electric wheelchair, and perform mouse-controlled computer functions would provide a measure of independence and greatly improved quality of life. Restoring useful leg movements is more difficult, since the major role of legs is to perform complex and precisely controlled movements during walking. Restoring control of all the necessary muscles would not restore proprioception—the sensation of the position of the legs in space—that is essential to normal and safe walking. The complexity of the problem notwithstanding, "Never say never" is an aphorism that definitely applies, since no one can fully predict what new developments both in knowledge of the nervous system and technology are over the horizon.

There are other developments that are perhaps closer to the horizon than a completely internal prosthetic brain-to-muscle circuit. Robotics simplifies the system by taking the limb out of the picture by using the electrical activity of the nervous system detected by implanted electrodes to manipulate a robotic arm. In the case of amputees, the robotic arm can be a prosthesis attached to the body. In individuals with spinal cord injuries the robotic arm could be an external device, possibly mounted on an electric wheelchair that could, for example, permit a quadriplegic individual to obtain food and drink from the refrigerator. A robotic arm of amazing dexterity and controllability has been developed by the U.S. Defense Advanced Research Projects Agency (Clark 2014). It is hard for me to imagine how a paralyzed arm fitted with electrodes could achieve comparable dexterity.

The body offers numerous opportunities for accessing internal electrical signals to control a robotic arm. The use of electrode arrays implanted into the motor cortex is one possibility, but in the case of amputees, brain surgery can be avoided by using the activity of extraneous muscles. The subject learns to flex muscles elsewhere in the body to signal to the prosthetic limb to make the desired movement. Electrodes detect the muscle activity and the computer algorithm directs the prosthetic limb to perform the intended movement.

Sending signals in the opposite direction, from the prosthetic limb to the primary somatosensory cortex, is also under investigation. The idea is to provide sensation from the prosthetic limb to the brain enabling the subject to feel, for example, the texture of objects detected by sensors on the prosthetic fingers. Interestingly, researchers do not necessarily need to have access to expensive prototype prosthetic arms to conduct this research; rhesus monkeys have been trained to operate virtual arms within a virtual reality environment displayed on a computer screen (O'Doherty et al. 2011).

BIONIC EARS

The production of bionic sensations presents different problems from the bionic control of movement. Sensations undergo progressive levels of processing as the signals they generate cross synapses and move deeper into the central nervous system. Therefore, the possibility of creating a useful and "lifelike" prosthetic sensation increases the closer the insertion point of a prosthetic sensation is to the site of input of the normal sensation, that is, to the receptors for the normal sensation. The most successful example to date is the cochlear implant, which produces bionic hearing by stimulating the axons in the auditory nerve in subjects who are profoundly deaf due to the loss of the receptor cells in the cochlea. The cochlea in the inner ear is an almost perfect setup for the insertion of bionic hearing into the nervous system.

Different sounds are different from one another because they are composed of sound waves of different frequencies. Pure tones have single frequencies, but most sounds are mixtures of tones of different frequencies. One might think it logical that a sound would be represented in the nervous system by trains of action potentials of the same frequency composition as the sound, but that is not what happens. The frequency of action potentials in the auditory axons encodes the loudness of the sound, not the frequency; soft sounds produce a low frequency of action potentials and loud sounds produce a high frequency of action potentials. Our perception of the pitch of a sound is determined by which axons in the auditory nerve are triggered to fire by the sound; that is, individual auditory neurons transmit sound of specific frequencies.

How the ear breaks down the sound vibrating the eardrum into its component frequencies and directs those frequencies to stimulate the appropriate auditory axon endings is a marvel of mechanical engineering. Sound is a pressure wave traveling in the air, and for the sound to be heard the pressure wave must be transferred from the air into the fluid in the cochlea. This involves the transfer of the sound across two membranes; the sound crosses the eardrum into the air-filled middle ear and then across the membrane of the oval window into the fluid-filled cochlea. Since there is air on both sides of the eardrum, vibrations arriving on the outside surface produce vibrations of the eardrum that are much stronger than would be produced if the eardrum had fluid against its inside surface. If you have ever had water get into the middle ear behind your eardrum you will have experienced the resulting muffling of the sound. Some mechanical advantage is needed to get sound into the fluid in the cochlea. Three ear bones (the hammer, anvil, and stirrup) spanning the middle ear provide lever action that concentrates the vibrations of the eardrum onto the smaller surface area of the membrane covering the oval window. In this way, the sound makes it across the air-to-fluid interface and enters the cochlea.

The cochlea consists of two parallel chambers, about 30 mm long, partitioned along their entire length by the basilar membrane. The chambers are not straight; they are arranged in a coil resembling a snail, but the coiling has little functional significance other than enabling the cochlea to fit nicely into the skull, so we can think of the cochlea as straight. After entering the first chamber of the cochlea through the oval window, the sound crosses the basilar membrane into the second chamber of the cochlea. The basilar membrane sorts the sound by frequency because it varies in stiffness along its length, which arises partly because the membrane is narrower near the oval window and becomes progressively wider toward the tip of the cochlea. The narrower regions of the membrane are stiffer, and the membrane stiffness decreases along the membrane as it widens toward the tip of the cochlea. The stiffer membrane regions preferentially vibrate at higher frequencies and the more compliant membrane regions preferentially vibrate at lower frequencies, so the frequency of the sound crossing the basilar membrane decreases with distance along the cochlea.

The receptors for sound called hair cells sit along the basilar membrane. Hair cells do not have hair, but they do have appendages called stereocilia, which are nothing like the head of hair that they resemble. Stereocilia are fingerlike extensions of the hair cell with cell membrane on their surface and a core of cytoplasm. The stereocilia extend up from the apical (top) end of the hair cell, contacting a membrane that runs parallel along the basilar membrane and called the tectorial membrane. (Note that the basilar membrane and the tectorial membrane are large structures, not cell membranes.) Thus, each hair cell extends between the two membranes with its base attached to the basilar membrane and its stereocilia contacting the tectorial membrane.

Vibration of the basilar membrane at the site of a hair cell causes movement between the basilar and tectorial membranes, which wiggles the stereocilia. The cell membranes at the tips of the stereocilia contain ion channels that are sensitive to the vibration. Like acetylcholine receptor channels in muscle, these channels pass small positive ions when opened. However, the extracellular fluid that bathes the membranes of the stereocilia is endolymph, which is not like ordinary extracellular fluid because it has a high concentration of potassium, and opening of the vibration-sensitive ion channels depolarizes the hair cells by allowing the entry of potassium. This causes the hair cells to release neurotransmitter, which binds to receptors on the auditory axon terminals, generating action potentials that travel to the brain to generate the sensation of sound. There is a round window covered with a membrane through which the sound leaves the second chamber of the cochlea, its task completed.

Think of the auditory axons innervating the hair cells along the basilar membrane as the strings of a piano, each tuned to bring into the brain sound of a frequency determined by its position along the keyboard, in this case along the basilar membrane. Cochlear implants play the piano by direct stimulation of the terminal axons of the cochlear nerve at their positions along the basilar membrane. Thus, the entire ear, up to and including the hair cells, is bypassed. Instead, an external device containing a microphone is positioned on top of the outer ear like an ordinary hearing aid. The microphone detects the sound and sends processed signals through a wire to a small broadcast antenna applied externally to the scalp. From there a wireless signal passes through the skull to the cochlear implant, which delivers the signal to an array of stimulating electrodes positioned along the basilar membrane, usually over a length of about 25 mm. The cochlea is about 30 mm long and the full low end of the frequency range is not quite reached by the electrode array. The electrode array is a single fiber containing about 20 internal electrodes of varying lengths that deliver their stimulating current at 20 positions spaced along the fiber. While the sorting of frequencies along the membrane produced by stimulation with the electrodes is relatively coarse grained compared to the natural activation of the cochlear axons by the hair cells, it is good enough to produce a useful experience of sound. Cochlear implants produce hearing with good speech discrimination, but there remain difficulties with the appreciation of music (Kohlberg et al. 2014), since the piano has only 20 keys.

BIONIC EYES

Vision, perhaps the most desirable target for the development of bionic sensation, presents formidable challenges since visual stimuli are complex and the site of normal input, the retina, is itself a computer. The retina is, in fact, an outgrowth of the cortex of the brain equipped with two types of photoreceptor cells—rods that detect the full range of wavelengths in the visible spectrum, and cones that detect color. Cone density is highest at the center of vision known as the fovea, and rod density is highest in a ring around the fovea. The density of both rods and cones drops toward the periphery of vision. The photoreceptor cells connect through circuitry formed by three types of neurons to the ganglion cells whose axons form the optic nerve that carries visual information to the brain (see Chapter 8). The available estimates of the number of photoreceptors and ganglion cells in the human eye vary somewhat, but as a ballpark estimate in each eye there are about 100 million photoreceptors, most of which are rods, conveying visual information to 1 million ganglion cells. The ganglion cell axons carry the information to the first stop in the brain, the lateral geniculate body. Further processing occurs there, and the geniculate axons then carry the message to the visual cortex located at the back of the brain, where further processing occurs.

It is hard to imagine creating an electronic eye that could produce output signals that represent vision and that could be interfaced at the level of the ganglion cells, geniculate neurons, or cortical neurons and restore sight, because these neurons all carry highly processed signals that are nothing like the pixels of the original picture detected by the photoreceptor cells. At the present stage of development restoring sight as we know it is not really the goal. Perhaps a better characterization is producing useful sensitivity to light. The ability to see shapes and track their movements would be a significant contribution to independence and quality of life for a completely blind individual. The first hint that this might be possible was the observation by Penfield and his colleagues that focal electrical stimulation of a point on the surface of the primary visual cortex produces the experience of seeing a spot of light in the visual field (Normann et al. 2009). Points of light elicited by electrical stimulation of the visual system are called phosphenes. Efforts to produce "vision" by stimulating the cortex began in the 1960s with the use of electrodes placed beneath the skull on the surface of the membranes covering the brain. Stimulation delivered via multiple electrodes produced patterns of phosphenes, but large currents were required that were close to the threshold for producing seizures.

The development of the technology to implant dense arrays of fine electrodes has produced renewed interest in the cortex as an interface for visual prostheses. Psychophysical investigations of normal humans into the question of how many pixels are required for a reasonably normal reading speed produced a surprise; subjects could read out loud at about 200 words per minute (about two-thirds of the reading aloud speed using normal vision) from a text displayed on only 625 pixels (Cha et al. 1992). This suggests that an array of 625 or so electrodes implanted into the primary visual cortex could produce useful "vision" that could restore reading and significantly aid mobility if each electrode elicited phosphenes independently. In some diseases that cause blindness and in many cases of blindness caused by accident the optic nerve is destroyed. The advantage of a camera-based bionic eye that directly stimulates the cortex is that the optic nerve is completely bypassed.

However, more progress has been made to date using the ganglion cells in the retina as the entry point to directly stimulate the optic nerve. Two types of bionic eyes have been developed that replace the photoreceptors in the retina (Ong and da Cruz 2012; Luo and da Cruz 2014). Both types electrically stimulate the ganglion cells to produce action potentials when light is detected. One type uses an external camera mounted in eyeglass frames as the light sensor, which sends signals wirelessly to an internal receiver connected to a grid of electrodes applied to the retina. The other type uses the cornea and lens of the eye itself instead of a camera. The implant is a sandwich in which an array of stimulating electrodes is connected point by point to an array of photodiodes, which is applied to the surface of the retina. Light entering through the cornea and lens of the eye is detected by the photodiodes, which deliver current to the electrodes that are localized at the position struck by the light, causing the ganglion cells there to fire action potentials. Thus, the optics of the eye is not replaced in this type of bionic eye, just the sensor, and the entire prosthesis is internal.

Bionic eyes can provide light sensitivity to individuals with blindness resulting from degeneration of the photoreceptor cells, such as retinitis pigmentosa, which otherwise leaves the internal circuitry of the retina and the ganglion cells intact. Electrical stimulation of the retina directly stimulates the ganglion cells to produce action potentials, bypassing for the most part any contribution of the other neurons in the retina. This is different from the hearing produced by cochlear implants. In normal hearing the sound detected by the hair cells is converted directly into action potentials in the auditory nerve, and further processing does not occur until the action potentials enter the brain, so no processing is bypassed. Thus, the "hearing" produced by cochlear implants is closer to normal than can be expected from the "vision" produced by a bionic eye where the neurons within the retina have been bypassed. Additionally, the pixel size in the bionic eyes currently under development is very much larger than in a natural eye, with each electrode activating the ganglion cells in an area that would have contained hundreds of photoreceptor cells, so the resolution is relatively low. Even so, one might imagine that some processes that take place in the retina might be taken over by the electronics of a future bionic eye. For example, individual ganglion cells have receptive fields with activating centers and inhibitory surrounds and vice versa (see Chapter 8). Although this arrangement, which serves to enhance the contrast between objects, is not likely to be reproducible by direct electrical stimulation of the ganglion cells, perhaps some compensation for the loss of contrast could be achieved by programming bionic eyes to deliver images with already enhanced contrast to the stimulating electrodes.

CONCLUSION

As with any complex issue it is possible to see both the potential and the many obstacles that must be overcome in order to interface the body and brain with machines that can compensate for losses due to injury or disease. But it must be appreciated that the exploration of the possible interfaces between our bodies and the machines we create is just beginning. One of the major messages of this book is that the solutions to complex problems spring from the broad base of human knowledge. Letting my imagination run wild, I can conceive of a system where electronic chips coated with neurotrophic factors that promote axon growth could be implanted into the brain or spinal cord, and axons would then grow onto and "innervate" the chip. Think of such a chip as a "keyboard" upon which the neurons whose axons connect to the chip can "type" a message that could be transmitted via a wireless transmitter to a computer screen or prosthetic device. I can further imagine the possibility that the axons of cortical motor neurons that have been severed by a spinal cord injury could regenerate a short distance to interface with such a chip connected to prosthetic circuitry that could deliver their messages to the spinal neurons that innervate the arms and legs, thus "curing" paralysis. Right now this is pure fantasy, but we are electric machines, and the full potential of our relationships with the electric machines of our creation is just beginning to be explored. The future of the human-machine relationship is most certainly beyond imagining.

EPILOGUE

To borrow a famous phrase from Charles Dickens, we do live in "the best of times and the worst of times." You can see it in many aspects of life in the twenty-first century. I feel it most keenly in the realm of advanced education. I graduated from high school in the United States in the mid-1960s and rode the wave of national commitment to higher education all the way to a Ph.D. from a prestigious university. One of the thrills of higher education for me was being around people who were actually engaged in discovering new knowledge. This was followed by a long career as a university professor with a funded research program. My four-year bachelor's degree from Rutgers University cost my working-class parents less than $10,000 total, room and board included, and my support for graduate work came from

National Institutes of Health and private-sector studentships—no student loans involved. This career path was not extraordinary in the 1960s and 1970s.

The problems today are well known. University educations have become far less affordable, and a Ph.D. in a biomedical research field and many other research fields hardly provides a good chance for a rewarding and financially secure future (see Alberts et al. 2014). Those who do make it through the substantial hurdles find a career landscape with a crushing amount of grant proposal writing and constant pressure to make discoveries that make money in the short term. Somehow the idea that research driven by curiosity will provide a base of knowledge and know-how, out of which the advances that help humanity and make money will grow, has lost favor. Now it seems that any research program that cannot manage to become a profit center is somehow a failure, contributions to human knowledge notwithstanding. I think that this cultural shift does not bode well for Western civilization.

The best of times in today's educational system arises from our tremendous access to information via the Internet. Universities play a huge role in this, from providing experts who write Wikipedia entries to providing online educational aids and massive open online courses (MOOCs) available to everyone with Internet access. Readers can Google any of the historical figures and most of the modern scientists and scientific terms mentioned in this book for instant access to additional information, definitions, and further reading. Do you want to observe how the Nernst equation and the Goldman-Hodgkin-Katz equation work? Google them and you will find a number of university websites that have computer simulations that calculate the membrane potential from the values for the ion concentrations and temperature that you key in. Are you interested in reading the original article in which Benjamin Franklin describes his kite-in-a-lightning-storm experiment? It is available online instantly and without charge from the Philosophical Transactions of the Royal Society website as are many of the historical works that report the great discoveries of modern science. Thus, the reach of universities now extends far beyond the direct impact on their own students.

Where do books fit into our future and, most relevant to this epilogue, where does this book fit in? My hope is that this book will spread an accurate—not dumbed down—knowledge of the basics of how the nervous system works to a population base much broader than just professionals in neuroscience. The book is intended to convey in a nutshell the picture of the function of the nervous system that I carry around with me after more than 40 years as a professional neuroscientist. My hope is that reading my book will stimulate the curiosity of many readers about how our bodies and minds and the bodies and minds of the other animals that share our planet with us really work. The book provides a perspective that can enhance the reader's understanding of information about the nervous system encountered in other books and in online sources. Since there is so much information out there, the perspective that is brought to bear is all-important for seeking out what makes sense and passing by what doesn't.

I believe that my book will help many readers realize that their nervous systems are wonderfully complex electrical machines that are constantly changing. Mental and physical activity can produce the growth of new connections with tremendously beneficial effects. There are important frontiers being explored in the areas of electronic prosthetics, but the nervous system that you have is an unparalleled achievement of human evolution. You need to take care of it.

I also hope that readers will have gained a better appreciation of our links to all of nature. Much of what we know about the basic workings of our nervous system comes from studies of frogs, squids, and a large variety of other animals. We need to have a reverence for nature both because of its wonders and because of what studying it has given us. Frogs are in deep trouble around the planet, and many species will likely go extinct. We need to give a little payback to the planet by figuring out how to save our environment, including the frogs.

I have intended my book to convey something about how scientific discovery works and the kinds of people who have made some of the great discoveries upon which our knowledge of the world, and in particular the nervous system, is based. Life is many things, and beyond all else it is a learning experience. The people who have looked at the world with the passion to figure out how things work have given us the modern world in which we now live. Much has been discovered through the efforts of unlikely people with creative minds and the willingness to go against authority.

Science has spawned many problems to be sure, but it is hard to imagine the alternative of living in "the demon-haunted world" described by Carl Sagan (1996). Our society needs to figure out how to maintain a vigorous effort in discovery research so that people with the passion and commitment to advance our knowledge can continue their quest and be there in universities to teach our young people and pass that passion on to them. Being taught exclusively by people whose only knowledge comes from reading about the discoveries made by others is just not the same.

What, specifically, are the major take-home messages of this book? Most general is the realization that we really are electric machines, since our sensations, movement, and everything in between, including consciousness, are electrical phenomena. My own reaction to this is a tremendous sense of wonder. We know this because of the amazing process of scientific discovery. So much has been learned through the development of technologies that bring the invisible into the visible world and experiments that bring the invisible world into sharp focus; for example, measuring the charge on a single electron and counting the number of synaptic vesicles released by an individual axon terminal. Many people know what Brownian motion is, but the realization that the random motions of atoms and molecules is strong enough to kick up a visible haze of dye particles in water creates an image of the world of atoms, ions, and molecules as one of constant, violent motion. This is the world of all the molecular machines that make our nervous systems work; for example, the voltage-gated sodium channels and acetylcholine receptor channels. It is this motion that makes chemical synaptic transmission so fast. It is the energy of this motion that is harnessed to produce the voltage across the cell membrane.

These take-home messages all support the major message of this book, which is how neurons produce and conduct electricity. I have made an effort to tell it like it really is in imagery that brings the mechanisms to life and does not require any prerequisite knowledge of biology, mathematics, or electrical circuits. How animals produce electricity has mystified and intrigued people for hundreds of years. It is most certainly worth knowing.

NOTES

REFERENCES

ACKNOWLEDGMENTS

INDEX

NOTES

CHAPTER 2. A WORLD OF CELLS, MOLECULES, AND ATOMS

1. Adapted from my favorite line in the movie *Braveheart*.
2. In scientific notation the decimal is placed after the first digit, giving 6.24, and the number of places that the decimal would have to be moved to the right to generate the actual number is expressed as an exponent of 10, in this case 10^{18}, which replaces quintillion. So the number of electrons or protons making up a coulomb of charge is written as 6.24×10^{18}, and it is verbalized as "six point two four times ten to the eighteenth power" (or just ". . . times ten to the eighteenth"). This makes sense since multiplying 6.24 by 10, 18 times, restores all the zeros giving the number that we started with. So 6.24×10^{18} can be used in calculations just as any ordinary number.

CHAPTER 3. THE ANIMAL BATTERY

1. The universal gas constant (R) times the absolute temperature in degrees Kelvin (T) is the amount of thermal energy in a mole of dissolved potassium, which powers the concentration gradient. F is Faraday's constant, which converts moles to coulombs of electric charge. To calculate the equilibrium potential for calcium (Ca^{2+}), which carries two positive charges, Faraday's constant would be multiplied by two. To calculate the equilibrium potential for a negative ion such as chloride, the concentration ratio would be flipped: $[Cl^-]_{in} / [Cl^-]_{out}$.

CHAPTER 6. HEART TO HEART

1. There are ambiguous cases, however. The cell bodies of somatosensory neurons located in ganglia along the spinal cord and elsewhere lack

recognizable dendrites. Instead, the axon splits into two branches a short distance from the cell body. One branch extends out into the body, innervating the sense organs, and the other branch extends into the spinal cord. Thus, the axon endings in the periphery serve the usual function of dendrites, receiving signals, in this case from the sense organs, and the direction of propagation of action potentials is toward the spinal cord. The cell body, off to the side, is not directly in the circuit.

2. It was once thought that all proteins are synthesized in the cell body, but it has been documented that messenger RNA coding for certain proteins is transported along the axons and directs protein synthesis in the axon terminals. This is an emerging field, but for now we will stick with the dogma, that the vast majority of axonal proteins are synthesized in the cell bodies, which is likely true under most conditions for the majority of proteins discussed in this book.

CHAPTER 7. NERVE TO MUSCLE

1. Described in Nicholls et al. 2001, 249.
2. Nanoscale dimensions are from Peters, Palay, and Webster 1976.
3. Quantification of neurotransmitter release and receptor binding is from Nicholls et al. 2001 and Tai et al. 2003.
4. Quantification is based on the analysis in Nicholls et al. 2001, 266.

CHAPTER 8. USE IT OR LOSE IT

1. Some values for the number of neurons in the cortex and brain published in books for a lay audience are wildly different. Edelman and Tononi (2000) claim 30 billion in the cortex.
2. Much of the discussion of LTP is based upon the treatment in Kandel et al. 2013.

REFERENCES

FURTHER READING

1. Animal Electricity

Smith, C. U. M., E. Frixione, S. Finger, and W. Clower. 2012. *The animal spirit doctrine and the origins of neurophysiology.* Oxford: Oxford University Press.

2. A World of Cells, Molecules, and Atoms

Coffey, P. 2008. *Cathedrals of science: The personalities and rivalries that made modern chemistry.* New York: Oxford University Press.

Johnson, G. 2008. *The ten most beautiful experiments.* New York: Vintage.

3. The Animal Battery

Alberts, B., A. Johnson, J. Lewis, M. Raff, K. Roberts, and P. Walter. 2002. *Molecular biology of the cell.* 4th ed. New York: Garland Science.

Kandel, E. R., J. H. Schwartz, T. M. Jessell, S. A. Siegelbaum, and A. J. Hudspeth, eds. 2013. *Principles of neural science.* New York: McGraw Hill.

Katz, B. 1966. *Nerve, muscle, and synapse.* New York: McGraw Hill.

Nicholls, J. G., A. R. Martin, B. G. Wallace, and P. A. Fuchs. 2001. *From neuron to brain.* Sunderland, MA: Sinauer Associates.

4. Hodgkin and Huxley before the War

Hodgkin, A. L. 1992. *Chance and design: Reminiscences of science in peace and war.* Cambridge: Cambridge University Press.

Katz, B. 1966. *Nerve, muscle, and synapse.* New York: McGraw Hill.

Smith, C. U. M., E. Frixione, S. Finger, and W. Clower. 2012. *The animal spirit doctrine and the origins of neurophysiology.* Oxford: Oxford University Press.

5. The Mystery of Nerve Conduction Explained

Hodgkin, A. L. 1992. *Chance and design: Reminiscences of science in peace and war.* Cambridge: Cambridge University Press.

Kandel, E. R., J. H. Schwartz, T. M. Jessell, S. A. Siegelbaum, and A. J. Hudspeth, eds. 2013. *Principles of neural science.* New York: McGraw Hill.

Katz, B. 1966. *Nerve, muscle, and synapse.* New York: McGraw Hill.

Nicholls, J. G., A. R. Martin, B. G. Wallace, and P. A. Fuchs. 2001. *From neuron to brain.* Sunderland, MA: Sinauer Associates.

6. Heart to Heart

Finger, S. 2000. *Minds behind the brain: A history of the pioneers and their discoveries.* Oxford: Oxford University Press.

Kandel, E. R., J. H. Schwartz, T. M. Jessell, S. A. Siegelbaum, and A. J. Hudspeth, eds. 2013. *Principles of neural science.* New York: McGraw Hill.

7. Nerve to Muscle

Katz, B. 1966. *Nerve, muscle, and synapse.* New York: McGraw Hill.

Nicholls, J. G., A. R. Martin, B. G. Wallace, and P. A. Fuchs. 2001. *From neuron to brain.* Sunderland, MA: Sinauer Associates.

8. Use It or Lose It

Kandel, E. R., J. H. Schwartz, T. M. Jessell, S. A. Siegelbaum, and A. J. Hudspeth, eds. 2013. *Principles of neural science.* New York: McGraw Hill.

Stroman, P. W. 2011. *Essentials of functional MRI.* Boca Raton, FL: CRC Press.

9. Broadcasting in the Volume Conductor

Ashcroft, F. 2012. *The spark of life: Electricity in the human body.* New York: W. W. Norton.

Finger, S., and M. Piccolino. 2011. *The shocking history of electric fishes.* New York: Oxford University Press.

Turkel, W. J. 2013. *Spark from the deep: How shocking experiments with strongly electric fish powered scientific discovery.* Baltimore, MD: Johns Hopkins University Press.

10. The Bionic Century

Ethier, C., E. R. Oby, M. J. Bauman, and L. E. Miller. 2012. Restoration of grasp following paralysis through brain-controlled stimulation of muscles. *Nature* 485:368–71.

Green, A. M., and J. F. Kalaska. 2010. Learning to move machines with the mind. *Trends in Neurosciences* 34:61–75.

Lebedev, M. A., and M. A. L. Nicolelis. 2006. Brain-machine interfaces: Past, present and future. *Trends in Neurosciences* 29:536–46.

Mudry, A., and M. Mills. 2013. The early history of the cochlear implant. *JAMA Otolaryngology–Head & Neck Surgery* 139:446–53.

Ong, J. M., and L. da Cruz. 2012. The bionic eye: A review. *Clinical and Experimental Ophthalmology* 40:6–17.

WORKS CITED

Alberts, B., A. Johnson, J. Lewis, M. Raff, K. Roberts, and P. Walter. 2002. *Molecular biology of the cell.* 4th ed. New York: Garland Science.

Alberts, B., M. W. Kirschner, S. Tilghman, and H. Varmus. 2014. Rescuing US biomedical research from its systemic flaws. *Proceedings of the National Academy of Science* 111:5773–77.

Amara, S. G., and M. J. Kuhar. 1993. Neurotransmitter transporters. *Annual Review of Neuroscience* 16:73–93.

Ashcroft, F. 2012. *The spark of life: Electricity in the human body.* New York: W. W. Norton.

Baker, C. I., E. Peli, N. Knouf, and N. G. Kanwisher. 2005. Reorganization of visual processing in macular degeneration. *Journal of Neuroscience* 25:614–18.

Becker, D. E. 2006. Fundamentals of electrocardiography interpretation. *Anesthesia Progress* 53:53–64.

Bennett, M. V. L., M. Wurzel, and H. Grundfest. 1961. The electrophysiology of electric organs of marine electric fishes: I. Properties of the electroplaques of *Torpedo nobiliana. Journal of General Physiology* 44:757–804.

Birks, R. I., and F. C. MacIntosh. 1957. Acetylcholine metabolism at nerve-endings. *British Medical Bulletin* 13:157–61.

Bittner, G. D., and D. Kennedy. 1970. Quantitative aspects of transmitter release. *Journal of Cell Biology* 47:585–92.

Bliss, T. V. P., and T. Lømo. 1973. Long-lasting potentiation of synaptic transmission in the dentate area of the anaesthetized rabbit following stimulation of the perforant path. *Journal of Physiology* 232:331–56.

Borck, C. 2008. Recording the brain at work: The visible, the readable, and the invisible in electroencephalography. *Journal of the History of the Neurosciences* 17:367–79.

Borgens, R. B., K. R. Robinson, J. W. Vanable, and M. E. McGinnis. 1989. *Electric fields in vertebrate repair.* New York: Alan R. Liss.

Brazier, M. A. B. 1984. *A history of neurophysiology in the 17th and 18th centuries: From concept to experiment.* New York: Raven Press.

Bryson, Bill, ed. 2010. *Seeing further: The story of science, discovery, and the genius of the Royal Society.* London: HarperCollins.

Campenot, R. B. 1977. Local control of neurite development by nerve growth factor. *Proceedings of the National Academy of Sciences USA* 74:4516–19.

———. 1982. Development of sympathetic neurons in compartmentalized cultures: II. Local control of neurite survival by nerve growth factor. *Developmental Biology* 93:13–21.

———. 1984. Inhibition of nerve fiber regeneration in cultured sympathetic neurons by local high potassium. *Brain Research* 293:159–63.

———. 1986. Retraction and degeneration of sympathetic neurites in response to locally elevated potassium. *Brain Research* 399:357–63.

Cha, K., K. W. Horch, R. A. Normann, and D. K. Boman. 1992. Reading speed with a pixelized vision system. *Journal of the Optical Society of America A* 9:673–77.

Cheng, A., and L. G. Tereshchenko. 2011. Evolutionary innovations in cardiac pacing. *Journal of Electrocardiology* 44:611–15.

Ciombor, D. M., and R. K. Aaron. 2005. The role of electrical stimulation in bone repair. *Foot and Ankle Clinics* 10:578–93.

Clark, L. 2014. Darpa's super-dexterous robot arm gets FDA approval. *Wired Magazine,* May 12. http://www.wired.co.uk/news/archive/2014-05/12/darpa-robotic-limb-fda-approval.

Cobb, M. 2002. Timeline: Exorcizing the animal spirits; Jan Swammerdam on nerve function. *Nature Reviews Neuroscience* 3:395–400.

Coffey, P. 2008. *Cathedrals of science: The personalities and rivalries that made modern chemistry.* New York: Oxford University Press.

Cole, K. S., and H. J. Curtis. 1939. Electric impedance of the squid giant axon during activity. *Journal of General Physiology* 22:649–70.

Connors, B. W., and M. A. Long. 2004. Electrical synapses in the mammalian brain. *Annual Review of Neuroscience* 27:393–418.

Copenhaver, B. P. 1991. A tale of two fishes: Magical objects in natural history from antiquity through the scientific revolution. *Journal of the History of Ideas* 52:373–98.

Curtis, H. J., and K. S. Cole. 1942. Membrane resting and action potentials from the squid giant axon. *Journal of Cellular and Comparative Physiology* 19:135–44.

Dale, H. 1936. Some recent extensions of the chemical transmission of the effects of nerve impulses. *Nobel Lecture.* http://www.nobelprize.org.

David, S., and A. J. Aguayo. 1981. Axonal elongation into peripheral nervous system "bridges" after central nervous system injury in adult rats. *Science* 214:931–33.

Davies, B. 1980. A web of naked fancies? *Physical Education* 15:57.

De Robertis, E. D. P., and H. S. Bennett. 1955. Some features of the submicroscopic morphology of synapses in frog and earthworm. *Journal of Biophysical and Biochemical Cytology* 1:47–63.

De Robertis, E., L. Salganicoff, L. M. Zieher, and G. Rodriguez de Lores Arnaiz. 1963. Acetylcholine and cholineacetylase content of synaptic vesicles. *Science* 140:300–301.

Edelman, G. M., and G. Tononi. 2000. *A universe of consciousness*. New York: Basic Books.

Ethier, C., E. R. Oby, M. J. Bauman, and L. E. Miller. 2012. Restoration of grasp following paralysis through brain-controlled stimulation of muscles. *Nature* 485:368–71.

Fatt, P., and B. Katz. 1951. An analysis of the end-plate potential recorded with an intra-cellular electrode. *Journal of Physiology* 115:320–70.

Feynman, R. P., R. B. Leighton, and M. Sands. 1963. *The Feynman lectures on physics*. Vol. 1. Reading, MA: Addison-Wesley.

———. 1964. *The Feynman lectures on physics*. Vol. 2. Reading, MA: Addison-Wesley.

Finger, S. 1994. *Origins of neuroscience: A history of explorations into brain function*. Oxford: Oxford University Press.

———. 2000. *Minds behind the brain: A history of the pioneers and their discoveries*. Oxford: Oxford University Press.

Finger, S., and M. Piccolino. 2011. *The shocking history of electric fishes*. New York: Oxford University Press.

Franklin, B. 1753. A letter of Benjamin Franklin, Esq; to Mr. Peter Collinson, F. R. S. concerning an electrical kite. *Philosophical Transactions of the Royal Society of London* 47:565–67.

Furshpan, E. J., and D. D. Potter. 1959. Transmission at the giant motor synapses of the crayfish. *Journal of Physiology* 145:289–325.

Garcia-Lopez, P., V. Garcia-Marin, and M. Freire. 2010. The histological slides and drawings of Cajal. *Frontiers in Neuroanatomy* 4:1–16.

Golgi, C. 1906. The neuron doctrine—theory and facts. *Nobel Lecture*. http://www.nobelprize.org.

Haberler, C., F. Alesch, P. R. Mazal, P. Pilz, K. Jellinger, M. M. Pinter, J. A. Hainfellner, and H. Budka. 2000. No tissue damage by chronic deep brain stimulation in Parkinson's disease. *Annals of Neurology* 48:372–76.

Harrison, R. G. 1910. The outgrowth of the nerve fiber as a mode of protoplasmic growth. *Journal of Experimental Zoology* 9:787–846.

Hebb, Donald O. 1949. *The organization of behavior: A neuropsychological theory*. New York: John Wiley & Sons.

Heuser, J. E., and T. S. Reese. 1973. Evidence for recycling of synaptic vesicle membrane during transmitter release at the frog neuromuscular junction. *Journal of Cell Biology* 57:315–44.

Heuser, J. E., T. S. Reese, M. J. Dennis, Y. Jan, L. Jan, and L. Evans. 1979. Synaptic vesicle exocytosis captured by quick freezing and correlated with quantal transmitter release. *Journal of Cell Biology* 81:275–300.

Hodgkin, A. L. 1937a. Evidence for electrical transmission in nerve: Part 1. *Journal of Physiology* 90:183–210.

———. 1937b. Evidence for electrical transmission in nerve: Part 2. *Journal of Physiology* 90:211–32.

———. 1947. The effect of potassium on the surface membrane of an isolated axon. *Journal of Physiology* 106:319–40.

———. 1963. The ionic basis of nervous conduction. *Nobel Lecture.* http://www.nobelprize.org.

———. 1992. *Chance and design: Reminiscences of science in peace and war.* Cambridge: Cambridge University Press.

Hodgkin, A. L., and A. F. Huxley. 1952a. The components of membrane conductance in the giant axon of *Loligo. Journal of Physiology* 116:473–96.

———. 1952b. Currents carried by sodium and potassium ions through the membrane of the giant axon of *Loligo. Journal of Physiology* 116:449–72.

———. 1952c. The dual effect of membrane potential on sodium conductance in the giant axon of *Loligo. Journal of Physiology* 116:497–506.

———. 1952d. A quantitative description of membrane current and its application to conduction and excitation in nerve. *Journal of Physiology* 117:500–544.

Hodgkin, A. L., and B. Katz. 1949. The effect of sodium ions on the electrical activity of giant axon of the squid. *Journal of Physiology* 108:37–77.

Hoffman, D. J. 2009. Fatal attractions: Curare-based arrow poisons, from medical innovation to lethal injection. Ph.D. diss., University of California, Berkeley.

Hubel, D. H. 1957. Tungsten microelectrode for recording from single units. *Science* 125:549–50.

———. 1981. Evolution of ideas on the primary visual cortex, 1955–1978: A biased historical account. *Nobel Lecture.* http://www.nobelprize.org.

Hubel, D. H., T. N. Wiesel, and S. LeVay. 1976. Functional architecture of area 17 in normal and monocularly deprived macaque monkeys. *Cold Spring Harbor Symposia on Quantitative Biology* 40:581–89.

Huxley, H. E. 1953. Electron microscope studies of the organisation of the filaments in striated muscle. *Biochimica et Biophysica Acta* 12:387–94.

Jaffe, L. F., and M. M. Poo. 1979. Neurites grow faster towards the cathode than the anode in a steady field. *Journal of Experimental Zoology* 209:115–28.

Johnson, G. 2008. *The ten most beautiful experiments.* New York: Vintage.

Kalmijn, A. J. 1982. Electric and magnetic field detection in elasmobranch fishes. *Science* 218:916–18.

Kandel, E. R., J. H. Schwartz, T. M. Jessell, S. A. Siegelbaum, and A. J. Hudspeth, eds. 2013. *Principles of neural science.* New York: McGraw Hill.

Katz, B. 1947. The effect of electrolyte deficiency on the rate of conduction in a single nerve fibre. *Journal of Physiology* 106:411–17.

Katz, B., and R. Miledi. 1972. The statistical nature of the acetylcholine potential and its molecular components. *Journal of Physiology* 224:665–99.

Keynes, R. 2005. J. Z. and the discovery of squid giant nerve fibres. *Journal of Experimental Biology* 208:179–80.

Kohlberg, G., J. B. Spitzer, D. M. AuD, and A. K. Lalwani. 2014. Does cochlear implantation restore music appreciation? *Laryngoscope* 124:587–88.

Köpfer, D. A., C. Song, T. Gruene, G. M. Sheldrick, U. Zachariae, and B. L. de Groot. 2014. Ion permeation in K⁺ channels occurs by direct Coulomb knock-on. *Nature* 346:352–55.

Lebedev, M. A., and M. A. L. Nicolelis. 2006. Brain-machine interfaces: Past, present and future. *Trends in Neurosciences* 29:536–46.

Levi-Montalcini, R. 1986. The nerve growth factor: Thirty-five years later. *Nobel Lecture*. http://www.nobelprize.org.

Loewi, O. 1936. The chemical transmission of nerve action. *Nobel Lecture*. http://www.nobelprize.org.

Lømo, T. 2003. The discovery of long-term potentiation. *Philosophical Transactions of the Royal Society of London B Biological Sciences* 358:617–20.

Lozano, A. M., and N. Lipsman. 2013. Probing and regulating dysfunctional circuits using deep brain stimulation. *Neuron* 77:406–24.

Lucas, K. 1917. *The conduction of the nervous impulse*. London: Longmans, Green.

Luo, Y. H.-L., and L. da Cruz. 2014. A review and update on the current status of retinal prostheses (bionic eye). *British Medical Bulletin* 109:31–44.

Millet, D. 2002. The origins of EEG. Abstract 24. Seventh Annual Meeting of the International Society for the History of the Neurosciences, Los Angeles. http://www.bri.ucla.edu/nha/ishn/ab24-2002.htm.

M'Kendrick, J. G. 1883. Notes on a simple form of Lippmann's capillary electrometer useful to physiologists. *Journal of Anatomy and Physiology* 17, part 3: 345–48.

Montgomery, D. 2012. *The rocks don't lie*. New York: W. W. Norton.

Morais-Cabral, J. H., Y. Zhou, and R. MacKinnon. 2001. Energetic optimization of ion conduction rate by the K⁺ selectivity filter. *Nature* 414:37–42.

Neher, E. 1991. Ion channels for communication between and within cells. *Nobel Lecture*. http://www.nobelprize.org.

Nicholls, J. G., A. R. Martin, B. G. Wallace, and P. A. Fuchs. 2001. *From neuron to brain*. Sunderland, MA: Sinauer Associates.

Normann, R. A., B. A. Greger, P. House, S. F. Romero, F. Pelayo, and E. Fernandez. 2009. Toward the development of a cortically based visual neuroprosthesis. *Journal of Neural Engineering* 6:1–8.

O'Doherty, J. E., M. A. Lebedev, P. J. Ifft, K. Z. Zhuang, S. Shokur, H. Bleuler, and M. A. L. Nicolelis. 2011. Active tactile exploration using a brain-machine-brain interface. *Nature* 479:228–31.

Ong, J. M., and L. da Cruz. 2012. The bionic eye: A review. *Clinical and Experimental Ophthalmology* 40:6–17.

Pereira, A. C., P. Aguilera, and A. A. Caputi. 2012. The active electrosensory range of *Gymnotus omarorum*. *Journal of Experimental Biology* 215:3266–80.

Perrin, J. B. 1926. Discontinuous structure of matter. *Nobel Lecture*. http://www.nobelprize.org.

Peters, A., S. L. Palay, and H. de F. Webster. 1976. *The fine structure of the nervous system: The neurons and supporting cells*. Philadelphia: W. B. Saunders.

Polikov, V. S., P. A. Tresco, and W. M. Reichert. 2005. Response of brain tissue to chronically implanted neural electrodes. *Journal of Neuroscience Methods* 148:1–18.

Robinson, K. R., and P. Cormie. 2007. Electric field effects on human spinal injury: Is there a basis in the *in vitro* studies? *Developmental Neurobiology* 68:274–80.

Roth, G., and U. Dicke. 2005. Evolution of the brain and intelligence. *Trends in Cognitive Sciences* 9:250–57.

Roux, F., and P. J. Uhlhaas. 2013. Working memory and neural oscillations: Alpha-gamma versus theta-gamma codes for distinct WM information? *Trends in Cognitive Sciences* 18:16–25.

Sagan, C. 1996. *The demon-haunted world.* New York: Ballantine.

Sakmann, B. 1991. Elementary steps in synaptic transmission revealed by currents through single ion channels. *Nobel Lecture.* http://www.nobelprize.org.

Shapiro, S., R. Borgens, R. Pascuzzi, K. Roos, M. Groff, S. Purvines, R. B. Rodgers, S. Hagy, and P. Nelson. 2005. Oscillating field stimulation for complete spinal cord injury in humans: A phase 1 trial. *Journal of Neurosurgery: Spine* 2:3–10.

Singer, S. J., and G. L. Nicholson. 1972. The fluid mosaic model of the structure of cell membranes. *Science* 175:720–31.

Smith, C. U. M., E. Frixione, S. Finger, and W. Clower. 2012. *The animal spirit doctrine and the origins of neurophysiology.* Oxford: Oxford University Press.

Tai, K., S. D. Bod, H. R. MacMillan, N. A. Baker, M. J. Hoist, and J. A. McCammon. 2003. Finite element simulations of acetylcholine diffusion in neuromuscular junctions. *Biophysical Journal* 84:2234–41.

Tansey, E. M. 2006. Henry Dale and the discovery of acetylcholine. *C. R. Biologies* 329:419–25.

Turkel, W. J. 2013. *Spark from the deep: How shocking experiments with strongly electric fish powered scientific discovery.* Baltimore, MD: Johns Hopkins University Press.

Vance, J. E., D. Pan, R. B. Campenot, M. Bussière, and D. E. Vance. 1994. Evidence that the major membrane lipids, except cholesterol, are made in axons of cultured rat sympathetic neurons. *Journal of Neurochemistry* 62:329–37.

Vesalius, A. 1543. *De humani corporis fabrica.* Basel: Ioannis Oporini.

Wallén, P., and T. L. Williams. 1984. Fictive locomotion in the lamprey spinal cord in vitro compared with swimming in the intact and spinal animal. *Journal of Physiology* 347:225–39.

Wheeler, J. A. 1979. A decade of permissiveness. *New York Review of Books,* May 17. http://www.nybooks.com/articles/archives/1979/may/17/a-decade-of-permissiveness.

Wiesel, T. N. 1981. The postnatal development of the visual cortex and the influence of environment. *Nobel Lecture.* http://www.nobelprize.org.

Zywietz, C. 2002. *A brief history of electrocardiography—progress through technology.* Hannover, Germany: Biosigna Institute for Biosignal Processing and Systems Research.

ACKNOWLEDGMENTS

I began writing this book on a sabbatical leave from the University of Alberta during the 2012–2013 academic year. I want to thank the Faculty Evaluation Committee of the Faculty of Medicine and Dentistry for the enthusiasm for my project that they conveyed when approving my sabbatical. This is my first book, and it was easy to wonder along the way whether it would ever see the light of day. My first glimpse of the end of the tunnel came when Dale Purves, a writer of many books in the field of neuroscience, responded with enthusiasm to an outline and writing sample that I had sent to him. That began a chain of events that led to acceptance of the book by Harvard University Press (HUP). I can't thank Dale enough for his support. I thank Michael Fisher, the editor at HUP when my book was accepted, and Thomas

Embree LeBien, who took over the editorial process when Michael retired. I learned much from my interactions with them. Most valuable was a four-hour discussion of the book with Michael over lunch at the University of Alberta Faculty Club. Many thanks to Lauren Esdaile, their editorial assistant, for all her attention and help. I thank the production editor, Christine Dahlin, and the copyeditor, Jennifer Shenk, for their excellent help producing the finished manuscript.

Many family members, friends, and colleagues have read and commented upon all or parts of the manuscript and offered their advice. Special thanks goes to my wife, Mary Kay Campenot, who read the entire manuscript, many parts more than once, and offered her excellent suggestions to improve the English and the clarity. Likewise, I thank my friends and scientific collaborators Jean Vance and Dennis Vance for their helpful reviews of a large part of the manuscript. I thank my friend and chair of the Department of Cell Biology, Rick Rachubinski, for his helpful comments on parts of the manuscript and for his support of this project and much else over many years. My thanks for their valuable contributions and support also go to Carl Amrhein, Sandra Blakeslee, Eric Campenot, Jennifer Campenot, Joel Dacks, Howard Howland, Sarah Hughes, Kendall James, David Kaplan, Blair Kopetski, Paul LaPointe, Claire Lavergne, Simon LeVay, Constance McKee, Lori Morinville, John Nicholls, Manijeh Pasdar, Thomas Simmen, Andrew Simmonds, John Teal, Susan Teal, and Carl Thulin.

INDEX

charge per mole, 67; charge transferred during a single action potential, 135

Chick embryo, 224–220, 285

Chloride: concentration gradient, 88; equilibrium potential, 88, 94, 182, 218–219; channels, 89–90, 93–95, 98, 153–155, 182, 218, 261; current, 94, 182, 218–220; permeability, 138, 218; role in inhibitory synapses, 182, 218–220

Chloride equilibrium potential, 88, 94, 102, 153, 182, 218–219, 261

Choline, 143, 207, 212

Choline acetyltransferase, 199–200

Circuit diagrams, 5, 20, 26–28, 35, 90, 92, 94, 116, 119, 141, 154, 219

Circuit model: of the cell membrane, 89–95; of a single location in squid axon membrane, 140–141

Cisternae, 205–206

Classical conditioning, 244–247

Clathryn, 206–207

Coated pits, 205–207

Cochlea: basilar membrane, 14, 305–307; implant, 14–15, 304, 306–307, 309

Cohen, Stanley, 226–227

Cold block, 115–121, 153

Cole, Kenneth (Kacy), 128–129, 132, 135, 138–139, 186

Collagen, 53, 73–74

Collagenase, 210

Competition in circuit formation, 228–229

Concentration gradient, 7–8, 67, 76–77, 82, 84–87, 94–95, 103, 118, 130, 194, 197

Conduction in frog nerve, 39–41

Conduction of the Nervous Impulse, The (Lucas), 110

Conduction velocity, 9, 123–124, 126–128, 150–153, 159–160, 274

Conformation, 78–80, 100–101

Connexon, 167, 273

Cortex: motor map, 298; sensory map, 298

Cortical neurons, 232–234, 241, 299–300, 308

Coulomb, 25

Crab axons: single axon recording, 122; reduced conduction velocity in oil, 123–124; overshooting action potential, 129–130; release of potassium during the action potential, 134–135

Cranial nerves, 223

Crayfish giant synapse, 167

Critic mentality in science, 169

Curare, 191, 197

Current, 22–30

Curtis, Howard, 128–129, 132, 135

Cutaneous dorsi muscle, 112

Cyclic adenosine monophosphate (cyclic AMP), 247–248, 251, 276–278

Cytoplasm, definition, 30

Dale, Henry, 175–181

Dalton, John, 53

Darwin, Charles, 266–267

David, Sam, 293, 295

Deadly nightshade, 178

Decremental conduction, 217

Defense Advanced Research Projects Agency, 303

De humani corporis fabrica (Vesalius), 49

Demon-Haunted World, The (Sagan), 282, 314

Dendrite: definition, 45; origin of the term, 165; role in integration of synaptic inputs, 182–183, 217–220; role of dendritic spines in learning, 250, 252–254

De Robertis, Eduardo, 199–200

Descartes, René, 46–48, 51, 105, 108, 198, 231

Development of the nervous system, 222–239

Dextrose, isotonic, 137

Diagrams, artistic license, 80

Diaminobenzidine, 204

Freezing point depression, 55

Friedrich Schiller University at Jena, Germany, 279

Frog heart, 10, 163, 177, 179, 181

Frog nerve-muscle: originated by Jan Swammerdam, 105–108; Hodgkin's experiments, 109–113; Lucas's and Adrian's experiments, 112–114

Frog skin, 283

GABA (gamma-aminobutyric acid), 182, 183, 218

GABA$_A$ receptor channels, 182, 218, 220

Galen, Claudius, 18–19, 21, 38, 40, 47–49, 161, 257

Galvani, Luigi, 39–41, 109, 114, 146, 161, 186, 191, 199, 213

Gamboge dye, 64–66

Gamma oscillations, 254

Ganglion cell, 15, 231–236, 307–310

Gap junction, 167

Gasser, Herbert, 122, 124

Gastrocnemius muscle, 109–110, 112, 117

Gating, mechanism of voltage-dependent, 78–80, 82, 98–103

Generator, static electric, 36–37

Gerard, Ralph W., 186–187

Glutamate, 182, 218, 247–248; involvement in LTP, 250–253

Glutamate receptors, 182, 183, 218, 219, 220; AMPA receptors, involvement in LTP, 251–253; NMDA receptors, involvement in LTP, 251–253

Goldman-Hodgkin-Katz equation, 88–89

Golgi apparatus, 173

Golgi, Camillo, 169

Golgi method, 9, 164–171

Gram-atomic mass, 66

Gram-molecular mass, 66

Growth cone, 168, 223

Grundfest, Harry, 124

Hair cell, 306

Hamburger, Viktor, 225, 228

Harrison, Ross Granville, 168

Harvard Medical School, 229

HCN channels, 274–278

Heart: muscle, 13, 108, 166, 175, 178, 214, 268, 270, 271, 273; pumping, 271–273; pacemaker, 273–278; purkenje fibers, 273, 278; action potential, 276–277

Heat theory of the elevator button, 20–22

Hebb, Donald, 243, 253–254

Heuser, John, 202–205

Hippocampus, 184, 216, 248–251, 253

Histologie du système nerveux de l'homme & des vertébrés (Ramón y Cajal), 169–170

Hodgkin, Alan, 3, 5, 8, 9, 88, 90, 95, 98, 105, 109–111, 113–124, 128–150, 153, 157, 159, 181, 186–187, 209, 213, 268, 312

Hodgkin and Huxley carrier model of sodium and potassium conductance, 144–145

Hooke, Robert, 50

Howland, Howard, 171

^{3}[H]-proline, 236–238

HRP (horseradish peroxidase), 204–205, 294

Hubel, David, 11, 229–230, 232, 234–235, 299

Huxley, Andrew, 3, 8–9, 88, 89, 95, 98, 105, 114, 118, 129–136, 139–150, 157, 181, 209, 213, 214

Huxley, Hugh, 213

Hydration shell, 69–70, 82, 84, 99

Hydrogen ion, 30, 56

Hydrophilic, 71–73

Hydrophobic, 71–73, 82, 144, 173

Île de Ré, 258

Imagery: importance to understanding, 4–5

Initial segment, 155, 182–183, 217–220

DATE DUE

			PRINTED IN U.S.A.